Toxic Histories

Toxic Histories combines social, scientific, medical and environmental history to demonstrate the critical importance of poison and pollution to colonial governance, scientific authority and public anxiety in India between the 1830s and 1950s. Against the background of India's 'poison culture' and periodic 'poison panics', David Arnold considers why many familiar substances came to be regarded under colonialism as dangerous poisons. As well as the criminal uses of poison, Toxic Histories shows how European and Indian scientists were instrumental in creating a distinctive system of forensic toxicology and medical jurisprudence designed for Indian needs and conditions, and how local as well as universal poison knowledge could serve constructive scientific and medical purposes. Arnold reflects on how the 'fear of a poisoned world' spilt over into concerns about contamination and pollution, giving ideas of toxicity a wider social and political significance that has continued into India's post-colonial era.

DAVID ARNOLD is Emeritus Professor of History at the University of Warwick. His published work includes *Colonizing the Body: State Medicine and Epidemic Disease in Nineteenth-Century India; Science, Technology and Medicine in Colonial India; Gandhi; The Tropics and the Traveling Gaze: India, Landscape, and Science, 1800–1856*; and *Everyday Technology: Machines and the Making of India's Modernity*.

Science in History

Edited by Simon J. Schaffer and James A. Secord

Science in History is a major series of ambitious books on the history of the sciences from the mid-eighteenth century through the mid-twentieth century, highlighting work that interprets the sciences from perspectives drawn from across the discipline of history. The focus on the major epoch of global economic, industrial and social transformations is intended to encourage the use of sophisticated historical models to make sense of the ways in which the sciences have developed and changed. The series encourages the exploration of a wide range of scientific traditions and the interrelations between them. It particularly welcomes work that takes seriously the material practices of the sciences and is broad in geographical scope.

Toxic Histories

Poison and Pollution in Modern India

David Arnold
University of Warwick

CAMBRIDGE
UNIVERSITY PRESS

CAMBRIDGE
UNIVERSITY PRESS

University Printing House, Cambridge CB2 8BS, United Kingdom

Cambridge University Press is part of the University of Cambridge.

It furthers the University's mission by disseminating knowledge in the pursuit of education, learning and research at the highest international levels of excellence.

www.cambridge.org
Information on this title: www.cambridge.org/9781107126978

© David Arnold 2016

This publication is in copyright. Subject to statutory exception and to the provisions of relevant collective licensing agreements, no reproduction of any part may take place without the written permission of Cambridge University Press.

First published 2016

Printed in the United Kingdom by Clays, St Ives plc

A catalogue record for this publication is available from the British Library

ISBN 978-1-107-12697-8 Hardback
ISBN 978-1-107-56573-9 Paperback

Cambridge University Press has no responsibility for the persistence or accuracy of URLs for external or third-party internet websites referred to in this publication, and does not guarantee that any content on such websites is, or will remain, accurate or appropriate.

Contents

Figures

Acknowledgements

This book owes its origins to the reading of two others. One is Alexandre Dumas' *The Count of Monte Cristo*, read one pleasant summer in the incongruous surroundings of Sag Harbor on Long Island, and the realization of how much of that fanciful tale of intrigue and enchantment related to ideas of Oriental toxicity and the Eastern poisoner. The other was Cecil Walsh's account of the Agra murders of 1911–12, which suggested much more even than the 'engrossing' poison story he unveiled. Since then, the book has grown into a more protracted engagement with the social, scientific, medical and environmental history of South Asia since the early nineteenth century, but over all this, to my mind, hangs the appalling tragedy of 'the world's greatest industrial disaster' – the poison gas leak at the Union Carbide plant in Bhopal in December 1984, even though it happened more than 30 years after Indian independence. This book is in part an attempt to provide a historical lineage for that catastrophe.

In the course of researching and writing this book, I have incurred many debts. I especially want to thank Jane and Bob Stine for their generous hospitability over the years, and also to express my indebtedness to the late Anil Agarwal for the inspiration and example of his environmental campaigning. My thanks go, too, to the many academic colleagues and friends who have regaled me with poison tales of their own, notably Clare Anderson, Michael Anderson, Barbara Dinham, Pratik Chakrabarti, Indira Chowdhury, Judy Farquhar, Margot Finn, Harald Fischer-Tiné, Marta Hanson, David Hardiman, Mark Harrison, Sarah Hodges, Shruti Kapila, Steve Legg, Jim Manor, Projit Mukharji, Henry Noltie, Robert Peckham, Margaret Pelling, Sanjay Sharma, Subhir Sinha, Claudia Stein, John Tresch and Michael Worboys. I am further indebted to the Wellcome Library, the British Library, the library of the School of Oriental and African Studies in London, the University of Warwick library, the National Archives of India in New Delhi and the Tamil Nadu State Archives in Chennai.

I further want to thank the organizers and participants of the various conferences, workshops and seminars at which earlier versions of these chapters were presented: at SOAS, King's College and the 'Reconfiguring the British' seminar in London; the Wellcome Unit for the history of Medicine in Oxford; the Swiss Federal Institute of Technology in Zurich; Johns Hopkins, Pennsylvania, Yale and Harvard in the United States; Academia Sinica in Taipei; the University of Hong Kong; and the National Centre for Biological Sciences in Bengaluru. An earlier version of Chapter 7 appeared as 'Pollution, Toxicity and Public Health in Metropolitan India, 1850–1939', *Journal of Historical Geography* 42 (2013): 124–33, and some of the themes of this book were presented in a preliminary form as 'The Politics of Poison: Empowerment and Subversion in Nineteenth-Century India', in David Hardiman and Projit Bihari Mukharji (eds), *Medical Marginality in South Asia: Situating Subaltern Therapeutics* (London: Routledge, 2012), 171–92. My thanks go, too, to Lucy Rhymer at CUP for her enthusiastic interest and helpful advice, to the anonymous reviewers of the manuscript and to Simon Schaffer and Jim Secord for welcoming the book into their Science in History series. My final and greatest word of gratitude is for Juliet Miller, without whose love and sustained support this book might never have reached its fruition.

Abbreviations

AR	*Asiatic Researches*
ARBSNC	*Annual Report of the Bengal Smoke-Nuisances Commission*
ARCE (NWP)	*Annual Report of the Chemical Examiner, North-Western Provinces*
ARCE (UP)	*Annual Report of the Chemical Examiner, United Provinces*
ARCED (Bengal)	*Annual Report of the Chemical Examiner's Department, Bengal*
ARMC	*Annual Report of the Municipal Commissioner*
CR	*Calcutta Review*
IAMS	*Indian Annals of Medical Science*
IJM	*Indian Journal of Malariology*
IJMR	*Indian Journal of Medical Research*
IMG	*Indian Medical Gazette*
IMS	Indian Medical Service
IOR	India Office Records
JASB	*Journal of the Asiatic Society of Bengal*
NAI	National Archives of India
NWP	North-Western Provinces
RCA (Bombay)	*Report of the Chemical Analyser, Bombay*
RCE (Punjab)	*Report of the Chemical Examiner, Punjab*
RHO	Report of the Health Officer
TMPSC	*Transactions of the Medical and Physical Society of Calcutta*
TNA	Tamil Nadu Archives
ToI	*Times of India*
UP	United Provinces

Introduction: poison traces

Poisoning is a universal phenomenon. Every society and every age has known it. Poisoning is present in the malignant guile and cunning deceit of the homicidal murderer, in the despairing fate of the man or woman driven to suicide, in the acute suffering and agonizing death caused by accidental contact with toxic substances. Poisoning is made manifest in the contamination and adulteration of food and drink, in the misuse of medicines and in the unintended effect of their toxic ingredients. In modern times, poisoning has resulted from toxic substances released into the environment through garbage tips, factory waste and industrial accidents, through stifling traffic fumes and the urban smog lowering over our cities. The stuff of fantasy, sensational murder trials, detective stories and 'true crime' tales, poisoning has caused the most public of tragedies, in which scores, even thousands, of people have perished: but it can also lurk, unseen and unsuspected, in our homes, workplaces and everyday lives. Poisons possess an incontestable materiality – in their chemical composition and physiological effects – but poison and poisoning are also extraordinarily rich in their semantic use and cultural deployment, as metaphors of malice and emblems of evil.

Modern society and modern science have conspired to create of poisons a kind of toxic globality. Toxic waste gets exported around the globe: it is dumped in deserts and landfill sites, in rivers and seas. Residual traces of toxic pesticides can be found in people, animals, plants and insects almost anywhere from the hot tropics to the frozen poles. Minerals, implicated for centuries in homicidal or accidental poisoning, are used in, or result from, industrial processing around the globe, their toxic detritus spilling over into cities and oceans. As commercial commodities, poisons are traded around the world and accidentally or inadvertently enter into an increasingly complex and interconnected domain of human ill health and ecological decay.[1] Forensic techniques and forms of

---

[1] Brett L. Walker, *Toxic Archipelago: A History of Industrial Disease in Japan* (Seattle: University of Washington Press, 2010).

toxicological knowledge that were once the preserve of the West are now the property of almost every country. Poisons, no less than the medicines they so often mimic, have proved highly mobile. We are familiar enough with culinary diasporas in which foods like sugar or maize have migrated from place to place, often, but not invariably, alongside the peoples among whom they originated. We could equally think of poison diasporas, in which toxic plants and minerals, poison practices and poison lore have migrated or been knowingly transferred from one region of the world to another, taking on a virtual universality of their own.[2]

And yet, for all this global trafficking and exchange, the claim can still be made that different societies have experienced and understood poison and poisoning differently from one another. The poisons used in suicide and murder have tended to be those most readily to hand – in the home and at the workplace – though the familiar and domestic nature of those substances might indicate the symbolic significance of their use, not merely their convenient availability.[3] Differently endowed by nature, history and culture, different societies have had different stores of poison (and different stories about poison) at their disposal. Mineral, vegetable and animal poisons found in one part of the world might, even now, be unknown or unobtainable in another.[4] Toxic diffusion has never entirely eliminated local poison cultures and local poison practices. Arjun Appadurai has observed that in a globalizing world the local can often reassert itself in surprising forms: poison might be one of the less pleasant of those surprises.[5] Apart from wide geographical variations in the distribution of poisons, the moral significance ascribed to poisoning, the degree of social acceptance or repugnance at its use, its historical associations and cultural resonances – each of these might still vary significantly from one society to another. What one set of people call a poison might

---

[2] For an example of the export of India's poison culture via a convict sent to Mauritius, see Angus Calder, *Gods, Mongrels and Demons* (London: Bloomsbury, 2003), 7–10. Conversely, another Indian poisoner was said to have learned his art from fellow convicts in Mauritius: A. H. Giles, 'Poisoners and Their Craft', *CR* 81 (1885): 110.

[3] On Britain, see Olive Anderson, *Suicide in Victorian and Edwardian England* (Oxford: Clarendon Press, 1987), ch. 10. One of the means by which unhappy wives in India sought to end their own lives, or those of their husbands, was by means of powdered glass, taken from their wrist bangles and crushed on a kitchen grindstone. Since for Hindu women bangles signified the married state, their breaking was symbolic of widowhood and separation. For one such case and its detection, see *RCA (Bombay), 1891*, 7.

[4] In 2009, a poisoning occurred in London involving a woman of South Asian origin and use of the drug aconite. In the subsequent trial, this poison was said to be extremely rare in Britain and therefore suggested a connection with India, where it was much more common: 'Curry Poisoning Woman Found Guilty of Murder', http://news.bbc.co.uk/1/hi/en gland/london/8492936.stm.

[5] Arjun Appadurai, *The Future as Cultural Fact: Essays on the Global Condition* (London: Verso, 2013), 4.

appear innocuous, even beneficial, to another. It is a cliché worth repeating that one man's meat is another man's poison. Poisoning is a universal phenomenon, but nowhere is its history the same.

Toxic Histories seeks to address in equal measure the social cognition and the scientific understanding of poisons and poisoning in nineteenth- and twentieth-century India. It aims to demonstrate, first of all, the existence within that spatial and temporal context of a significant and distinctive poison culture, one that was of sufficient importance and extent to generate an extensive literature and critical commentary all its own. The book sets out to show how poison and poisoning were socially and culturally embedded in pre-colonial India, and how India's experience of colonialism (until 1858 under the English East India Company and thereafter under the British Crown until partition and independence in 1947) helped transform the social parameters and political understanding of local poison practice. An existing poison culture became politicized and polemicized under British rule, but Indian agency also had a part in that process of transformation. *Toxic Histories* seeks to demonstrate the place of science, and the social authority of science, in colonial modes of poison governance, to show how toxicology – the scientific study of poisons and poisoning – became a salient and indicative part of the way in which science spoke to both an imperial and an Indian public.

On the face of it, the case for India having a prominent and distinctive culture of poison and poisoning might appear unpromising. If one turns to the statistical record, the medium through which so much of the colonial understanding of India was presented, poisoning might seem to be of only marginal significance. In the great tally of human sickness and mortality in British India, poisoning was dwarfed by such monumental killers as smallpox, cholera, malaria and plague. In 1873, in the vast and populous North-Western Provinces, out of 765,534 patients treated in dispensaries and allied medical institutions, only 1,377 cases (0.2 per cent) involved poisoning, and of these only 56 proved fatal.[6] In 1903, in the same province, now renamed the United Provinces, out of more than 3 million individuals attending hospitals and dispensaries, barely 5,000 were poison cases.[7] In 1911, there were some 7,000 poison cases and a mere 38 fatalities.[8] At a time when tens of thousands of people across this and other Indian provinces were dying of famine and epidemic disease, remarkably few cases of poisoning – homicidal, suicidal or accidental – were recorded. Even allowing for substantial under-reporting (especially

---

[6] *Annual Report of the Dispensaries of the North-Western Provinces, 1865*, 7, 10.
[7] *Annual Report of the Hospitals and Charitable Institutions of the United Provinces, 1903*, 69A.
[8] *Annual Report of the Civil Hospitals and Dispensaries of the United Provinces, 1911*, 43A, 51A.

of non-fatal cases), statistically speaking, poisoning appears as little more than a minor footnote to a far more compelling history of mass mortality.[9]

Similarly, if one looks at the provincial police reports of the time, poison cases do not figure particularly prominently compared to other categories of murder and assault or the high level of property crimes and public order offences. In 1911, the police in the United Provinces took up sixty-six poison cases (twenty-one of them attributed to 'professional poisoners') and secured only seventeen convictions. Significantly, though, a further eighty-two cases involved cattle poisoning – in India, people were not poison's only victims.[10] If one turns to the annals of Indian criminality in general, the activities of the thugs, who deceived, strangled and dismembered travellers on the highways of nineteenth-century India, attracted wide publicity and impelled the colonial state to adopt drastic measures for their suppression. Dacoit gangs and 'criminal tribes' provoked further exceptional, often draconian, measures. By contrast, poisoning and poisoners do not appear to have caused comparable levels of public distress and state concern.[11] But, if that were so, and if poisoning were of little material significance, what are we to make of the observation made by Bengal's chemical examiner in 1902 that 'no country in the world furnishes anything like the amount of toxicological material that India does'?[12] Was this seemingly scientific utterance mere prejudice and fantasy? If the historiography of Indian criminality has largely ignored poisoning, does that lacuna connote an actual absence or a flaw in the writing of that history?

Nor does the extensive historiography of health, medicine and disease in British India offer a more promising platform for an enquiry. Academic scholarship has in the main been committed to the idea that the history of medicine is concerned with disease on the one hand and healing on the other. It has not overly concerned itself with the indeterminate middle ground, with the 'constitutive ambivalence' of the *pharmakon*, in which substances that function as medicines serve also as poisons or operate as both poisons and medicines simultaneously.[13] Monographs and general surveys of colonial medicine have appeared in which the complexities of

---

[9] Ira Klein, 'Death in India, 1871–1921', *Journal of Asian Studies* 22 (1973): 639–59; David Arnold, *Colonizing the Body: State Medicine and Epidemic Disease in Nineteenth-Century India* (Berkeley: University of California Press, 1993).

[10] *Report on the Administration of the Police of the United Provinces, 1911*, 11.

[11] Poisoning receives only passing mention in Radhika Singha, *A Despotism of Law: Crime and Justice in Early Colonial India* (Delhi: Oxford University Press, 1998), and still less in Mark Brown, *Penal Power and Colonial Rule* (Abingdon: Routledge, 2014).

[12] C. H. Bedford, 'Notes on Some Toxicological Experiences in Bengal and in the Punjab', *IMG* 37 (1902): 202.

[13] On the 'constitutive ambivalence' of poisons, see Frédéric Obringer, *L'Aconit et L'Orpiment: Drogues et Poisons en Chine Ancienne et Médiévale* (Paris: Librairie Arthème Fayard, 1997), 12–14.

health provision under British rule have been meticulously examined and extensively critiqued or in which the changing configuration of India's 'indigenous systems of medicine' has been exposed to close and careful scrutiny. And yet, in all this extensive and painstaking literature, poisons and poisoning seldom rate even a mention. But does it make sense to discuss medicine without giving due consideration, too, to poison, medicine's evil twin and toxic other? Might not the diverse systems of medicine in nineteenth- and twentieth-century India be described with equal validity as systems of toxicological – as well as therapeutic – knowledge? The social use and political profile of narcotic drugs like opium and cannabis have been examined at length in recent scholarship, but their role as deliberate or accidental poisons has drawn little comment.[14] Administrative action and legislative control over various forms of intoxication – from alcohol to opium – has been noted.[15] But, by contrast, the landmark legislation of the Indian Poisons Act of 1904 has been ignored, though a sceptic might still want to argue that the fact of this enactment coming decades after similar laws in Britain might itself be taken as evidence of the secondary importance of 'poison scares' in India.[16] There never was a high-profile enquiry in India into toxicity in food and drink to match the Royal Commission on Arsenical Poisoning in Britain in 1901.[17] Should one therefore conclude that the free availability and criminal use of arsenic – the quintessence of nineteenth-century homicidal and accidental poison – was never a substantive issue in India, never of such importance as to warrant public alarm and impel state action?

Where poison appears at all in India's medical history, it is as an aside, as a flawed miasmatic concept, in which ill health was understood, before the bacteriological revelations of the late nineteenth century, as occasioned by poison-like emanations from swamps, jungles, graveyards and overcrowded habitations. Thus conceived, poison represents little more than a misconceived aetiology, an epiphenomenon attributable to the

---

[14] James H. Mills, 'Drugs, Consumption, and Supply in Asia: The Case of Cocaine in Colonial India, c. 1900 c. 1930', *Journal of Asian Studies* 66 (2007): 345–62; Richard Newman, 'Early British Encounters with the Indian Opium Eater', in James H. Mills and Patricia Barton (eds), *Drugs and Empires: Essays in Modern Imperialism and Intoxication, c. 1500–c. 1930* (Basingstoke: Palgrave Macmillan, 2007), 57–72.

[15] Harald Fischer-Tiné and Jana Tschurenev (eds), *A History of Alcohol and Drugs in Modern South Asia: Intoxicating Affairs* (London: Routledge, 2014).

[16] In Britain, the sale of poisons was controlled by an Act to Regulate the Sale of Arsenic in 1851, followed by two Pharmacy Acts in 1852 and 1868: James C. Whorton, *The Arsenic Century: How Victorian Britain Was Poisoned at Home, Work, and Play* (Oxford: Oxford University Press, 2010), ch. 5.

[17] *Royal Commission on Arsenical Poisoning: First Report of the Royal Commission Appointed to Inquire into Arsenical Poisoning from the Consumption of Beer and Other Articles of Food or Drink: Part I: Report* (London: His Majesty's Stationery Office, 1901).

medical uncertainties of a barely scientific age, a metaphorical substitute for the still-elusive materiality of microbes and 'germs'.[18] In ways suggestive of poison's local configuration, one of the few routes by which poisoning has entered current writing on the history of medicine and science in British India has been through non-human agency and impact – through snakes and their venom and the criminal poisoning of cattle. Both suggest the need for a more complex understanding of poisoning in animal and environmental, as well as human, histories.[19]

It is, then, the task of this book to make the case that poisons and poisoning *were* of practical importance and ideological significance to science, governance and society in British India. *Toxic Histories* seeks to show that a multi-layered but also evolving understanding of poisons and poisoning existed in India between the early 1830s and the late 1940s. It found expression not just in toxicology – a scientific domain in its own right – but also in related fields such as medicine and public health, botany, chemistry, ethnology and criminology.

Readers might reasonably question whether proposing such a strong cultural and ultimately political connection between poisons and India might not smack of Orientalism. It might seem to suggest that there was something dangerous, odd and atavistic about poisons and their usages in India that might not apply to poisons and poisoning in, say, ancient Rome, Renaissance Italy or Victorian Britain – to name just three times and places where poisoning had a significant social presence or in whose histories it has a powerful imaginative hold.[20] There is, indeed, much to connect India with the recurrent literary and artistic trope of the treacherous, guileful and malicious 'Oriental poisoner', or of the Orientalized European who deploys an Eastern knowledge of poisoning against his or her enemies.[21] Poison might speak with authority and passion to a Western sense of Eastern enchantment and danger – a connection evident even in that ur-Orientalist text, *The Arabian Nights*, with its tales of drug potions and books with poisoned pages. An association with the Orient and with India haunts not just Thomas De Quincey's opium-fuelled

---

[18] On miasmatic 'poisoning', see Mark Harrison, *Climates and Constitutions: Health, Race, Environment and British Imperialism in India, 1600–1850* (New Delhi: Oxford University Press, 1999), 26–57.

[19] Pratik Chakrabarti, *Bacteriology in British India: Laboratory Medicine and the Tropics* (Rochester, NY: Rochester University Press, 2012), ch. 4; Saurabh Mishra, *Beastly Encounters of the Raj: Livelihoods, Livestock and Veterinary Health in North India, 1790–1920* (Manchester: Manchester University Press, 2015), ch. 6.

[20] See Ian Burney, *Poison, Detection, and the Victorian Imagination* (Manchester: Manchester University Press, 2006), ch. 1. On Orientalism and its imaginative powers, see Edward W. Said, *Orientalism* (London: Routledge & Kegan Paul, 1978).

[21] The phrase 'The Eastern Poisoner' appears as a chapter heading in H. L. Adam, *The Indian Criminal* (London: John Milne, 1909).

fantasies, but also his thoughts on the 'fine art' of murder.[22] In the 1840s, the French novelist Alexandre Dumas devoted a chapter of *The Count of Monte Cristo* to the subject of toxicology, in which Madame de Villefort is instructed in the use of poisons. Throughout the book, just as opulence is identified with the legendary gold and diamond mines of Gujarat and Golconda, so is the art of poisoning intricately bound up with the Orient in general and India in particular.[23] In Léo Delibes' romantic opera *Lakmé*, first performed in 1883, the heroine, daughter of a Brahmin priest, dies after chewing a leaf from a datura (thorn apple) bush.[24] As late as 1924, one of the inmates at the sanatorium in Thomas Mann's novel *The Magic Mountain* is an eccentric Dutchman, who acquired his knowledge of strychnine during a visit to India's Coromandel Coast and uses it to end his own life.[25] Few traits so characterize the India of the imperial era as this apparently Oriental appetite for poison.[26] But to posit a connection, as this book does, between India and poisons is not thereby to endorse a fantastical or derogatory stereotype, or even simply to investigate the 'truth' behind such a deceptive facade. Rather, it is necessary, in Homi Bhabha's words, to 'recognize the stereotype as an ambivalent mode of power and knowledge', to interrogate its usages and effects and to examine how the imagining of, and fantasizing about, poison and poisoning became interwoven with a broader narrative of science and society.[27]

Meanings and concepts

It may be helpful at the outset to say a word about what the terms 'poison', 'pollution' and 'toxicity' mean and the uses to which they are put in this book. It is no more possible for colonial India than for Britain or anywhere else in the nineteenth century to offer a definitive answer to the seemingly simple question: 'What is a poison?' One plausible answer might lie in the authoritative texts that helped establish the modern science of poisons in Europe and so informed investigation into poisons in British India. In his foundational treatise on toxicology, published in French in 1815 and in English a year later, the chemist Mathieu Orfila stated: 'The name of poison is given to any substance, which, taken inwardly in a very small

---

[22] Thomas De Quincey, *Confessions of an English Opium Eater* (London: Penguin, 1971); idem, *On Murder* (Oxford: Oxford University Press, 2006).

[23] Alexandre Dumas, *The Count of Monte Cristo* (London: Penguin, 2003), ch. 52.

[24] Earl of Harewood (ed.), *Kobbé's Complete Opera Book* (London: Putnam, 1981), 821–26.

[25] Thomas Mann, *The Magic Mountain* (London: Vintage Books, 1999), 548, 578–79.

[26] For a contrary attempt to downplay the perceived exceptionality of poisoning in India, see Giles, 'Poisoners', 78–122.

[27] Homi K. Bhabha, *The Location of Culture* (London: Routledge, 1994), 66.

dose, or simply applied in any kind of manner to a living body, depraves the health, or entirely destroys life.' He went on to identify different categories of poisons – from 'stupefying and narcotic poisons' like opium and datura, through mineral and metallic poisons like arsenic, mercury and copper, to 'septic, or putrefying poisons', such as arose from 'contagious miasmata, emanating from pestiferous bodies, or bales of merchandize coming from a place infected with the plague'.[28] Even nineteenth-century commentators found Orfila's poison typology vague and excessively wide ranging. A leading British toxicologist and authority on medical jurisprudence, Alfred Swaine Taylor, proposed a more concise definition. Poison was any substance which, 'when administered in small quantity, is capable of acting deleteriously on the body'. This, though, he conceded, might define poison too narrowly. Even substances like common salt could become poisonous if consumed in large quantities, and not all poisons (such as snake venom) could be said to be 'taken internally'. In common usage, Taylor noted, to speak of a poison generally signified a 'deadly poison', an intensely toxic substance like arsenic or strychnine that could rapidly destroy human life and not just inflict temporary discomfort. This, he suggested, was also much closer to what toxicologists themselves had in mind.[29]

Modern dictionary definitions follow broadly the same lines as those proposed by Taylor more than a century and a half ago, but with some significant additions. Typically poison is identified as 'any substance which, taken into *or formed in the body*, destroys life or impairs health' – thus recognizing the capacity of the body itself to produce poisons or toxins.[30] A recent medical encyclopaedia notes that many common substances can prove poisonous to the human body when taken in excess but prefers to reserve the term 'poison' for materials that are 'harmful in small quantities'. Even so, the list of poisons is still a very long one, and includes 'practically all [medicinal] drugs and many minerals and synthetic substances'.[31] We also now know more than Orfila and Taylor did about how poisons actually work. Corrosive poisons, such as acids, alkalis and many disinfectants, act by altering the chemical state of proteins in the body and so cause 'indiscriminate damage' to living matter. Other poisons interfere with the body's chemical reactions – cyanides, for

---

[28] M. P. Orfila, *A General System of Toxicology, or Treatise on Poisons* (London: E. Cox & Son, 1816), 1–12.

[29] Alfred S. Taylor, *On Poisons, in Relation to Medical Jurisprudence and Medicine* (London: John Churchill, 1848). On the definitional problem faced by Taylor and his contemporaries, see Burney, *Poison*, 57–60.

[30] *Chambers English Dictionary* (Cambridge: Chambers, 1988), 1127 (emphasis added).

[31] Peter Wingate with Richard Wingate, *The Penguin Medical Encyclopedia* (3rd ed., London: Penguin, 1988), 380.

instance, prevent the transfer of oxygen in living cells and induce 'chemical suffocation'. Vegetable poisons (which can be more toxic in small doses than mineral poisons) are often chemically related to substances in the body, which they displace in vital reactions. 'Their use as [therapeutic] drugs depends on these effects', and so poisoning is 'an exaggeration of the medicinal effect'.[32]

The law did not always view poisoning in the same light as medical science. The Indian Penal Code, dating from 1860, and the Criminal Procedure Code that followed it gave no specific definition of poison. There was, however, general recognition of another of Taylor's medico-legal axioms – that a poison was 'a substance which, when absorbed into the blood, is capable of seriously affecting health or of destroying life'.[33] For the purposes of India's criminal courts, there was no need to define what constituted a poison because 'any act done with the intention of causing injury, no matter by what means caused', constituted a punishable offence. If that act resulted in the victim's death, then the offence became murder or culpable homicide. If it merely caused injury, it amounted to 'simple hurt' or 'grievous hurt', according to the degree of harm sustained. In other words, intention counted for more than method, or, in the case of poison, the nature of substance and the size of the dose administered.[34] Such a legalistic approach was of little help, however, in defining what constituted poison or in differentiating between various poisons and their toxic effects.[35] Even the Indian Poisons Act of 1904 did not define poison but instead drew up a schedule of poisons that fell under the scope of the act. In practice, poison in British India came to be understood through a series of interrelated cognitive processes, executive strategies and scientific techniques. These included diverse modes of scientific investigation, notably a botanical-medical route through which poisons were identified via their plant origins, social uses and physiological effects, and a biochemical-bacteriological route that involved laboratory testing to establish the presence of a specific chemical or bacterial agent. These, in turn, helped fuel a system of medical jurisprudence to which ethnology and criminology were also added. Just as botany and chemistry sought to make visible otherwise secretive modes of poisoning, so medical techniques (such as the use of post-mortems)

---

[32] Ibid.

[33] Cited in Patrick Hehir, *Opium: Its Physical, Moral, and Social Effects* (London: Ballière, Tindall & Cox, 1894), 358.

[34] Ibid, 358–59; L. A. Waddell, *Lyon's Medical Jurisprudence for India* (5th ed., Calcutta: Thacker, Spink, 1914), 415.

[35] There remained many anomalies: for instance, powdered glass, when used in suicide and murder, was classed as a poison, though its action was not chemical but that of a 'mechanical irritant': Waddell, *Lyon's*, 514.

provided other means by which to make poisons and poisoning legible, detectable and so amenable to science and the exercise of state power. These often highly technical modes of enquiry exemplified the view that an understanding of poison, even when scientifically grounded and tested, in actuality reflected the concept of poison less as an objective reality and discreet materiality than as a politically contingent and socially constructed idea.[36] To a significant degree, a poison was a poison not just because of its plant origins or chemical properties but because it had long been used as such or because it had, over time, become a signifier for specific cultural traits and social characteristics.

There is heuristic value, moreover, in recruiting the idea of poisons and poisoning to inform and substantiate a still wider notion of toxicity. It is part of the ambition of this work to present toxicity as an overarching concept, and not merely as a set of disparate ideas and practices, to see how such a concept emerged, attained authority and evolved alongside other key concepts of the period – such as poverty and development – to help construct and appraise India's modernity.[37] Although dictionary definitions make little distinction between what is toxic and what is poisonous, the value for the present work of using both terms conjointly is that they make it possible to trace the evolution of ideas of poisons from being substances external to the body to having a living presence within the body: as 'poisons' in the increasingly redundant miasmatic sense passed out of use, they became reconstituted in a bacteriological age as 'toxins', as poisonous elements generated or active within the body. Physicians began to talk, too, about the 'toxic' side effects of chemotherapy, of the adverse but unavoidable consequences of using ancient poisons like arsenic and antimony to modern medicinal effect. Between the late nineteenth century and the early decades of the twentieth, medicine constructed a new language, a new conceptualization, of what toxicity might mean – for the body, for medicine, for society.

In recent years, 'toxicity' has taken on a new semantic significance as a means of conceptualizing and critiquing a poisoning not just of people but of the environment at large, as modern industry and urban living have polluted (i.e., brought toxicity to) food, water and atmosphere, and had a wide-ranging and detrimental impact on soils, plants and animals. As William Cronon has put it, in describing 'the pain of a poisoned world', since the mid-twentieth century we have become aware of 'the proliferating presence of toxic compounds in the webs of ecological relationships

---

[36] On contructivism, see Jan Golinski, *Making Natural Knowledge: Constructivism and the History of Science* (Cambridge: Cambridge University Press, 1998).
[37] Reinhart Koselleck, *The Practice of Conceptual History: Timing History, Spacing Concepts* (Stanford: Stanford University Press, 2002).

that sustain life on the planet'. We live in 'a new age of toxicity'.[38] The classic – and most influential – statement of this has been Rachel Carson's *Silent Spring*, first published in 1962, in which modern chemical pesticides are equated with 'poison' and the present becomes, nightmarishly, 'an age of poisons'.[39] In one striking passage, Carson writes:

> Our attitude towards poisons has undergone a subtle change. Once they were kept in containers marked with skull and crossbones; the infrequent occasions of their use were marked with utmost care that they should come in contact with the target and with nothing else. With the development of the new organic insecticides and the abundance of surplus planes after the Second World War, all this was forgotten. Although today's poisons are more dangerous than any known before, they have amazingly become something to be showered down indiscriminately from the skies. Not only the target insect or plant, but anything – human or non-human – within range of the chemical fallout may know the sinister touch of the poison.[40]

Silent Spring has great relevance for the present discussion in which a conscious endeavour is made to demonstrate for India since the early nineteenth century a vital connectivity between poisoning and pollution as dual manifestations of a shared toxic principle. Pollution is no more easily defined than poison. There is the added difficulty that in India the term 'pollution' has been applied both to ritual states of impurity, along with the contaminating substances and contacts that cause it, and to the release of harmful chemical and organic substances into the environment. Despite their difference in meaning and context, both notions of pollution are closely interrelated, as when poor, low-caste Hindus are seen as being materially the source of environmental pollution as well as being ritually impure and likely to transmit that impurity to others. To speak of 'toxicity' is thus to posit a series of interconnected material states and cultural concepts that range from the calculated homicidal and suicidal use of a known poison to the perception and effects of discharging a polluting substance into a watercourse or the atmosphere. Likewise, contamination, adulteration, putrefaction and intoxication all lurk on the fringes of this toxic continuum. Whatever the intentionality (or apparent lack of it) that might lie behind such actions and effects, in all these varied registers of poisoning and pollution, there is the suggestion of something that is not just harmful but abnormal and abhorrent: they are acts against nature, however that nature and its presumed purity happen to be construed.[41]

---

[38] William Cronon, 'Foreword: The Pain of a Poisoned World', in Walker, *Toxic Archipelago*, ix.

[39] Rachel Carson, *Silent Spring* (London: Hamish Hamilton, 1963), 143.   [40] Ibid., 127.

[41] This understanding of toxicity comes close to (without replicating) Mary Douglas's formulation about dirt being matter 'out of place'. Mary Douglas, *Purity and Danger: An Analysis of Concepts of Pollution and Taboo* (London: Routledge, 2002), 44.

Carson's seminal work has made it possible for more recent scholars like Lawrence Buell to identify the emergence of a 'toxic discourse' and the 'fear of a poisoned world'.[42] The pastoral idyll about which so much was once written and fondly imagined (as in the landscape literature of pre-industrial America) has morphed into spectral visions of an 'environmental apocalypse', in which nature stands impoverished, degraded and denuded, testament to moral and social degeneration as well as environmental decay.[43] This book does not seek for India so apocalyptic a vision of past and present, but it does aspire to present a darker, more cautionary, narrative of India's progression to modernity and an alternative reading to histories of science, medicine and society that incline towards technological triumphalism and the luminous vision, favoured in recent years by the Hindu right, of India as a 'shining' land.

Among the many environmental disasters and toxic calamities that have come to epitomize the scenario of a 'poisoned world' – the Chernobyl nuclear accident in 1986, the Exxon Valdez oil spill in 1989, the Fukushima disaster of 2011 – the poison gas leak at the Union Carbide chemical plant at Bhopal in India on the night of 2–3 December 1984 is the one most pertinent to the present discussion. At least 3,800 people died within hours of toxic gas leaking from the plant, with as many as 10,000 deaths following within days and between 15,000 and 20,000 deaths over the next two decades. Bhopal – a calamity that involved poisoning and pollution on a colossal scale – has become synonymous with industrial disaster and environmental catastrophe.[44] So, while Buell and Carson are primarily concerned with North America, there are good reasons – including Bhopal – to import the idea of a toxic discourse into the history of South Asia, where studies of the environment have concentrated on the rural environment and have, until lately, neglected urban and industrial India or seen urbanization and industrialization as having little bearing on environmental history until after 1947.[45] Employing a language of toxicity makes it possible to posit a connectivity between nineteenth-century poison scares and murder cases and more recent developments in which poison and pollution have taken

[42] Lawrence Buell, 'Toxic Discourse', *Critical Inquiry* 24 (1998): 639.

[43] Ibid., 642, 648; Lawrence Buell, *The Environmental Imagination: Thoreau, Nature Writing, and the Formation of American Culture* (Cambridge: Belknap Press, 1995), ch. 9.

[44] Edward Broughton, 'The Bhopal Disaster and Its Aftermath: A Review', *Environmental Health* 4 (2005), at www.ehjournal.net/content/4/1/6.

[45] On urban environmental history, see Michael Mann, 'Delhi's Belly: On the Management of Water, Sewage and Excreta in a Changing Urban Environment during the Nineteenth Century', *Studies in History* 23 (2007): 1–31; Awadhendra Sharan, *In the City, Out of Place: Nuisance, Pollution, and Dwelling in Delhi, c. 1850–2000* (New Delhi: Oxford University Press, 2014).

on a far wider environmental and social significance. A focus on poison helps redefine the environmental, as well as the scientific and medical, history of South Asia.

Toxic histories

There are several strands to the argument of this book. Chapter 1 begins the exploration of the interrelated questions of knowledge, power and difference that run throughout the book. It does so in the first instance by considering poisons in relation to the 'social life of things' and outlines poison knowledge and poison use in pre-colonial India and the early decades of British rule. While all societies can be said to have their own poison histories and cultures, India had a long acquaintance and distinctive relationship with poisons at both a material and mythic level, and from elite culture to the lives of the poor. What colonial epistemology came by the mid-nineteenth century to regard as poisons were substances already employed in India for a great diversity of purposes. Vegetable and mineral poisons were incorporated into aphrodisiacs and elixirs or used as components in compound medicines: they were conceived of as stimulating or 'heating' substances that energized the body's organs and countered impotence and lethargy. Poison knowledge was deeply rooted in medical tradition and physiological understanding. But toxic substances were also implicated in the commission of robberies and murders, in suicide, homicide, abortion and infanticide. Poison further presented itself to human experience through the perils posed by venomous snakes, while poisoning – accidental or unavoidable – was a danger connected, too, to the dietary hazards of the poor, who in times of hardship and famine resorted to 'surrogate' foods with high levels of toxicity. Poison – in its various manifestations – was a matter of almost everyday acquaintance.

The second chapter explores the nature and meaning ascribed to poison substances as they became subject to colonial investigation and incorporated into British India's scientific practice and ideological apparatus. In part this was an enquiry, predicated on colonial needs and concerns, into 'native' usages, an aspect of a wider process of 'bioprospecting' that involved 'useful' plants as well as dangerous substances. Given the botanical underpinning to colonial medicine, the primary focus was on vegetable poisons, with aconite, datura and opium singled out for critical attention: discussion of arsenic, the fourth main poison, is held over to Chapter 6. While recognizing their different 'lives' and characters, colonial enquiry tended to qualify or contest the therapeutic value hitherto assigned to these drugs and to stress the dangers posed by their medicinal use and criminal deployment. This was science thinking politically, 'like a state', rather than from a position of impartiality.[46] The sustained critique of India's poison culture was empowering to the colonial regime in that it supported claims made about Indian barbarism and the superiority of Western science and civilization. Although practitioners of indigenous medicine were rebuked for their negligent or reckless use of toxic drugs, the emphasis given to the inherent dangers of the Indian pharmacopoeia paradoxically created a space for Indian agency and authority. The colonial critique enabled Indians in colonial service to contest this negativity, utilizing their poison expertise to facilitate the exchange of toxic, as well as therapeutic, knowledge between 'rival' medical systems.[47]

It is in the nature of poison to unsettle, to call into question everyday assumptions about love and loyalty, security and the established order. Chapter 3 illustrates one aspect of this destabilizing effect: 'poison panics' and their association with groups of real or supposed poisoners. Such scares among Indians mostly pertained to fears about poisoned or contaminated food and water. But the colonial order was itself not immune to comparable moments of alarm in which sections of the Indian population were categorized as poisoners. The examples discussed include poisoning by Indian princes and of datura *thugi* (the drugging and robbing of travellers) in the light of what C. A. Bayly has identified as the colonial 'information order' and its deficiencies.[48] Until the 1890s, the state did not regard either of these specific threats as sufficiently dangerous to require new legislation or dedicated policing measures. However, as Chapter 4 goes on to show, one of the ways in which the colonial regime responded to these panics and other poison-related challenges was by seeking a more effective knowledge of its colonial subjects and their poison practices. This was done primarily through the techniques of an evolving toxicological science but supplemented by a form of medical jurisprudence devised to suit India's social as well as physical circumstances. The state's main toxicological agency, the provincial chemical examiners, played a crucial part in the identification and detection of poison cases, and their reports, and the stories they tell about science, provide a rich source for this study. But both the practical limitations as well as the ideological biases of colonial toxicology and medical

---

[46] 'Thinking like a state' draws on James C. Scott, *Seeing Like a State: How Certain Schemes to Improve the Human Condition Have Failed* (New Haven: Yale University Press, 1998).

[47] John C. Hume, 'Rival Traditions: Western Medicine and Yunan-i Tibb in the Punjab, 1849–1889', *Bulletin of the History of Medicine* 51 (1977): 214–31.

[48] C. A. Bayly, *Empire and Information: Intelligence Gathering and Social Communication in India, 1780–1870* (Cambridge: Cambridge University Press, 1996).

jurisprudence raise further questions about the authority of science and the reliability of evidence in poison cases.

In Chapter 5, attention turns to the place of intimacy in poison practices, and the importance of private lives and domestic locations as 'spaces of experience' in which poisoning so frequently occurred. Part of the value of this emphasis on intimacy is as a means of showing how poison cases provided a vital connection between private and public, exposing the intimate details of private lives to public scrutiny, censure and sensationalist reporting. One of the ways in which poison – and the science that sustained enquiry into its criminal use – acquired a public audience and a modern identity was through the dissemination of 'poison tales'. Collectively and cumulatively, accounts of criminal investigations and poison trials helped inform stereotypical representations of the 'Oriental poisoner' or of poisoning as an Oriental contagion that threatened European lives. In narrating European vulnerability and Indian duplicity, poison tales spoke primarily to an imperial public, but poison stories were not for European consumption alone. They – or others like them – also attracted Indian audiences and posed moral dilemmas of their own for an Indian public.

Having looked previously at vegetable poisons and their scientific investigation, the sixth chapter turns to arsenic, its growing availability and use in nineteenth-century India and its implication in criminal activity, including homicide and cattle poisoning. As in the West (but significantly later), arsenic poisoning provided the primary impetus for legislation – the Indian Poisons Act of 1904 – to control the sale of toxic substances. Although the act was the outcome of decades of debate about the need for more effective poison governance, the act also signalled an important stage in India's 'toxic transition' – the shift away from a state-centred preoccupation with person-specific, poison-related crime to the growing concerns of a middle-class Indian public and with the poisoning that resulted not from homicidal intent but from adulterated foods and unregulated drugs sold in the bazaars. Poisons – and the scientific knowledge associated with them – became an Indian public cause, not just an imperial concern.

The following chapter expands on these issues by focusing on urban India and evolving ideas of the city's poisoned or polluted environments. It shows how concerns about the environment and its 'poisoned' nature had a long history connected, in the first instance, to miasmatic theories of tropical disease but persisting, in a revised form, in medical ideas and public health practices well into the twentieth century. The tropics bred 'poison' but poisons were also needed to bring tropical toxicity under control. In cities like Bombay and Calcutta (now known as Mumbai and

Kolkata, respectively), measures were adopted to regulate 'nuisances' caused by urban trades, developing industries and the growing contamination of air, water and food, but the growth of modern industry and the increasing use of hazardous commodities like lead and arsenic threatened health in the home and in the workplace. While the middle classes vented their concern over environmental hazards that seemed to affect their health and well-being, it was commonly the poor who were held responsible for causing or exacerbating public nuisances. By the 1930s, various forms of inspection and control had been established, notably with respect to smoke pollution, but paucity of funds, the smallness of the inspectorate and the difficulty of prosecuting offenders militated against their effectiveness. The concluding section of the chapter reflects on the long-term significance of poison for post-independence India.

It might be possible, in the final analysis, to argue for toxicity as representing a kind of Manichaean divide between good medicine and bad medicine.[49] The history of poisons and poisoning might be said to contain little that is unequivocally good. It can be seen – often is seen – as a history of unmitigated evil, a deplorable sequence of reckless conduct, malevolent desire and destructive effect. Perhaps, remembering Bhopal, this book ultimately supports such a negative prognosis. But it also seeks to engage, from a historical perspective, with the more positive outcomes that emerged from the social understanding and scientific exploration of poison and pollution – the stimulus they provided, for instance, for the local investigation of the chemical properties and social uses of various toxic substances, the incentive poison and pollution provided for an investigation (however hesitant) into the lives and livelihoods of the rural and urban poor and the agency they gave to Indians in science and the public sphere. One significant, if neglected, ingredient in the science of British India involved learning from toxicity, harnessing its constructive possibilities and believing that poisons could be regulated and redeployed for positive ends, not least in the service of the nation and the cause of development. And yet one cannot fail to see, too, how a kind of scientific hubris, an overconfident belief in the ability to contain and manage toxicity, or, perhaps worse still, a blindness to its presence and effects, could in the end prove to be a Faustian pact from which only hardship, suffering and destruction would result.

---

[49] For instances of poisoning as 'bad medicine', see David Wootton, *Bad Medicine: Doctors Doing Harm since Hippocrates* (Oxford: Oxford University Press, 2006).

1   The social life of poisons

Poisons are material objects. For much of history they have been commodities, much like any others, that are bought, sold and exchanged, and
have served a wide range of commercial, industrial, medicinal and domestic uses. Some poisons may be rare and exotic, but most are common
substances. They belong to a history of 'everyday things'.[1] As such,
poisons help us to appreciate the nature and texture of people's lives,
the material circumstances of their daily existence. They are registers of
experience – of the poverty and desperation, for instance, that might drive
an individual to murder or to suicide. As material substances, poisons
have 'social lives'. They exist within a social matrix that endows them with
specific meanings and distinct usages. Their cultural signification is
informed by history, religion and myth, and their status and character
are influenced by local, as well as global, configurations of race, class and
gender.[2] Later chapters of this book are concerned with the public face of
poisons, with the politicization and governance of poisonous substances.
The function of this chapter, however, is to establish as a baseline the
quotidian existence of poisons in Indian society in the nineteenth century,
to show how and why poisons were, in the main, everyday objects.

India's poison culture

Poisoning has always had public roles and political uses. From the death
of Socrates, condemned to drink hemlock in Athens in 399 BCE, through
the poison-laced intrigues of the Borgias in Renaissance Italy, to the
murder of the Russian defector Alexander Litvinenko in London in

[1] Daniel Roche, *A History of Everyday Things: The Birth of Consumption in France, 1600–1800*
(Cambridge: Cambridge University Press, 2000), 1–7.
[2] Arjun Appadurai (ed.), *The Social Life of Things: Commodities in Cultural Perspective*
(Cambridge: Cambridge University Press, 1986), especially Igor Kopytoff, 'The
Cultural Biography of Things: Commoditization as Process', ibid., 64–91. Cf. Susan
Reynolds Whyte, Sjaak van der Geest and Anita Harden, *Social Lives of Medicine*
(Cambridge: Cambridge University Press, 2002); Zheng Yangwen, *The Social Life of
Opium in China* (Cambridge: Cambridge University Press, 2005), 1–2.

2006, poisoning has been deployed as an instrument of state – to remove rivals and dissidents, to eliminate individuals whose popularity had grown too great, to punish those suspected of betrayal and treachery. India has a long history, often inseparable from myth, of poison practices. One example from India's antiquity was the attempt on the life of the Mauryan king Chandragupta, a contemporary of Alexander the Great, in the fourth century BCE, by means of a 'poison maiden'.[3] This designation was sometimes applied to women who administered poison to their victim concealed in food or drink. More dramatically, the phrase signified women whose bodies had, over time, through small but incremental doses, become so impregnated with poison as to be fatal to any man who lay with them or even came into close physical contact. Such toxic embraces formed part of Indian legend and yet the idea was considered sufficiently plausible to be included in Sanskrit medical treatises like the *Sushruta Samhita*, dating back to well before the start of the Common Era.[4] There was more than a hint of misogyny and male anxiety in this idea. As one later commentator observed of Sushruta's account of the *vishakanya* or 'venomous virgin', 'If she touches you, her sweat can kill. If you make love to her, your penis drops off like a ripe fruit from its stalk.'[5] Clearly, such women were to be feared and shunned: but the idea of poisoning as a politics of bodily intimacy and betrayal, as something to be evaded but also perhaps knowingly embraced, is one that will recur throughout this book.

A rich vein of poison-lore runs, too, through the medieval and early modern chronicles of Islamic South Asia from the early thirteenth century onwards.[6] These relate how poison was concealed in the food and drink offered by Hindu 'unbelievers' to destroy Muslim conquerors, or how the wise ruler and canny commander preserved his life by suspecting and detecting a poison plot against him. Poison became a formulaic device by which the warrior and statesman was tested and thereby revealed his superior powers.[7] In his memoirs, the Mughal ruler Babur recounted how in 1526, the year of his conquest of northern India, he suspected

---

[3] L. A. Waddell, *Lyon's Medical Jurisprudence for India* (5th ed., Calcutta: Thacker, Spink, 1914), 414.

[4] Jivanji Jamshedji Modi, 'The Vish Kanya or Poison Damsel of Ancient India', in *Anthropological Papers, Part IV* (Bombay: British India Press, 1929), 226–39; Dominik Wujastyk, *The Roots of Ayurveda: Selections from Sanskrit Medical Writings* (New Delhi: Penguin, 2001), 124–26.

[5] Cited in Wujastyk, *Roots*, 126.

[6] Peter Jackson, *The Delhi Sultanate: A Political and Military History* (Cambridge: Cambridge University Press, 1999), 77, 176.

[7] H. M. Elliot and John Dowson, *The History of India, as Told by Its Own Historians: The Muhammadan Period, Vol. II* (London: Trübner, 1869), 522–23.

an attempt had been made to poison him. The food served to him made him sick, so he fed his vomit to a dog (a common means of detecting poison): it became lethargic and 'out of sorts' but did not die.[8] According to contemporary European travellers, India's Mughal rulers used poison to eliminate rivals and fulfil their dynastic aims. Emperors bestowed on princes and the high-ranking nobility robes of honour, known as *khilats*. Their conferment normally signalled the high regard in which the ruler held the recipient, so they could hardly be refused. But such a gesture of honour and appreciation was sometimes inverted: the lining of the robes was impregnated with poison, causing the wearer to die a miserable, agonizing death. Aurangzeb, last of the 'great Mughals', who died in 1707, was particularly fond of this device, once unsuccessfully attempting to eliminate his rebellious son Akbar with a 'killer *khilat*'.[9]

Reports from the Mughal era (again mostly from European sources) further indicate how the emperors used poison to silence or punish dissent. The appeal of this toxic instrument of power was that it could be used without recourse to the shedding of blood by decapitation or by maiming (a fate more fitting for a common criminal) or to imprisonment and exile, from which there was always the possibility of return. Poisoning offered a less violent, arguably less extreme, method of execution, one that allowed the victim to preserve a degree of dignity and status. Like Socrates sipping his hemlock, high-ranking subjects suspected of treachery were obliged to drink a concoction of raw opium known as *post*: this caused stupor, madness and, without a reprieve, death.[10] François Bernier noted in his description of Aurangzeb's reign:

This *poust* is nothing but poppy-heads crushed, and allowed to soak for a night in water. This is the potion generally given to Princes confined in the fortress of Goüaleor [Gwalior in central India], whose heads the Monarch is deterred by prudential reasons from taking off. A large cup of this beverage is brought to them early in the morning and they are not given anything to eat until it be swallowed ... This drink emaciates the wretched victims, who lose their strength and intellect by slow degrees, become torpid and senseless, and at length die.[11]

[8] [Babur], *Babur Nama: Journal of Emperor Babur* (New Delhi: Penguin, 2006), 285–88.
[9] Michelle Maskiell and Adrienne Mayor, 'Killer Khilats, Part 1: Legends of Poisoned "Robes of Honour" in India', *Folklore* 112 (2001): 23–45, esp. 32. Further references to 'poison robes' and political poisonings can be found in James Tod, *Annals and Antiquities of Rajasthan* (3 vols, London: Oxford University Press, 1920).
[10] John Fryer, *A New Account of East India and Persia in Eight Letters* (London: Richard Chiswell, 1698), 32, refers to *post* as consisting of bhang (Indian hemp or cannabis) and datura, 'the deadliest sort of Solanium, or Nightshade', which made the recipient 'foolishly mad'. For *post* and its effects, see R. N. Chopra and N. N. Ghose, 'Addiction to "Post" – Unlanced Capsules of *Papaver somniferum*: Part II', *IJMR* 19 (1931): 415–21.
[11] François Bernier, *Travels in the Mughal Empire, A. D. 1656–1668* (Westminster: Archibald Constable, 1891), 105–07. *Post* was reserved for members of the Mughal

Political execution by means of poison might appear to elevate the practice of poisoning into the realms of the exceptional, and yet even these tales of courtly conspiracy and imperial punishment hint at a more subaltern consciousness. They are an indication of how widely in pre-colonial India knowledge existed about the nature and effects of vegetable and mineral poisons. These accounts passed from imperial memoirs and court chronicles into bardic tales, common legend and popular folklore. Further, the circulation of Indian poison tales in the West, disseminated through widely read narratives like Bernier's, helped shape European perceptions of India as a land of poisons and poisoners. Such tales might even eventually fuel the anxieties of the colonial British, concerned about their own power and vulnerability.[12]

These are themes to which later chapters of this book will return. But for the moment let us persist with the idea that one of the things that made poisons and poisoning matters of everyday importance, that gave them 'lives', was their pervasiveness – or at least the pervasive idea of their presence – in the cultural life of India and in the intimacies of imagination, myth, religion and speech that poison engendered or enriched. Unlike the Judaeo-Christian tradition, in which poison tales are conspicuously absent, Hindu mythology contains a remarkably rich store of poison-lore.[13] The central Creation myth of the Churning of the Ocean involved a cosmic struggle between gods and anti-gods, between good and evil. It eventuated in the simultaneous generation of the nectar of immortality (*amrita*) and a fiery poison (*visha*) that threatened to engulf and destroy the entire universe. From this primordial source all the world's poisons were said to originate.[14] The conceptualization of poison as a universal negative, a coruscating, life-destroying principle in dialectical opposition to the nectar of purity and virtue, coursed through Indian religious imagery and permeated idioms of the everyday. In another epic myth, the god Shiva swallowed (and forever retained in his throat) the poison spewed out by the serpent Vasuki, thus preventing the contamination of the ocean of milk, which would have destroyed the gods and burned the universe to cinders. Such was the ferocity of this venom that Shiva was himself only saved from destruction by his consort Parvati, who seized him by the neck to prevent his ingesting the poison, but the venom,

royal family 'as being a more secret death, free from the outward signs of laying violent hands upon one of the Blood Royal': ibid., 107.

[12] Michelle Maskiell and Adrienne Mayor, 'Killer Khilats, Part 2: Imperial Collecting of Poison Dress Legends in India', *Folklore* 112 (2001): 164–65.

[13] British writers saw closer parallels with Rome than with the Bible: A. H. Giles, 'Poisoners and Their Craft', *CR* 81 (1885): 84–85.

[14] N. Subramanya Aiyar, 'Certain Facts Regarding the Poison-Lore of the Hindus', *IMG* 31 (1896): 6.

though contained, remained so intense that it turned his body blue.[15] In further renditions of this story, the terrifying spectre of the *visha purusha*, or poison monster, created by the Churning of the Ocean could only be defeated by the intervention of Brahma, Lord of Creation, who forced the monster to shed its terrifying form and take refuge in the lesser form of venomous snakes, scorpions and spiders.[16] Again, in one of the many tales told of the young Krishna, an ogress named Putana tried to poison him with the milk from her breasts, but she was killed when the infant god, unharmed, sucked the life out of her.[17] There is some resemblance here – in the identification of poisoning as a female trait perversely allied to sex, reproduction and nurture – with the 'poison maiden' myth cited earlier.[18]

Poison thus stood as the sign for all manner of things that were wicked, corrupting and destructive but also, more positively, those substances that tested – and thereby proved – the divine attributes and doughty qualities of gods, saints and heroes. Monarchs, holy men and their humbler imitators might thus demonstrate their prowess by eating poison or proving immune to its effects. In the Vedas, the god Rudra was described as 'the master of poison and medicines', a controller of toxic substances as well as of *soma* and other consciousness-altering drugs.[19] At the other end of the spectrum of divinity, poison might be used to test whether an old woman was a witch – even though death alone might be the demonstration of her innocence.[20] Poison thus bore a dual identity: it was both destroyer and enabler. It could cause or threaten the destruction of kings, gods, demons, even the universe, but it could also be the means by which – contained, overcome, redirected – the superior intellect, courage, wisdom or spiritual strength of the god, sage and warrior-king could find demonstration and proof. This was a mythic principle, but it was also an idea fecund in its historical implications, in its social and even scientific application.

Poison featured in the high Hindu tradition and in Sanskrit texts but it surfaced, too, in many of the popular, vernacular works of the *bhakti* (devotional) canon. Thus, for the ardent devotee, separation from the divine presence could itself be understood as a species of poison, so

[15] Veronica Ions, *Indian Mythology* (London: Newnes, 1983), 43.

[16] K. M. Shyam Sundar, *Treatment for Poisons in Traditional Medicine* (Madras: Centre for Indian Knowledge Systems, 1996), 3.

[17] Wendy Doniger, *The Hindus: An Alternative History* (New York: Penguin, 2009), 478.

[18] Lee Siegal, *Sacred and Profane Dimensions of Love in Indian Traditions as Exemplified in the Gitagovinda of Jayadeva* (Delhi: Oxford University Press, 1978), 130–31.

[19] Doniger, *Hindus*, 120.

[20] 'It is a common superstition in the country that witches withstand the action of poisonous drugs': *ARCED (Bengal), 1899*, 16.

intense was the sense of physical and spiritual loss.[21] In the *Adi Granth*, the collection of sacred hymns compiled during the period of the Sikh gurus (1469–1708), poison is repeatedly invoked as a metaphor to represent the falsehoods and temptations that stand in the way of the devotee's pursuit of the divine, just as egotism and neglect of the guru amount to poison in almost literal form. In the verses of the 'Siri Raag', the deluded worshipper, swayed by passion and by lust ('the great poison'), imagines himself in pursuit of nectar when he is in fact drawn to the poison of ignorance and so lured to moral and spiritual death.[22] In another of the hymns, the 'Raag Gauri', the devotee who is attached to 'lust, wrath and love', 'eats poison considering it a sweet thing', while of the false believer it is said that 'In his heart is poison, [but] with his mouth he utters nectar'.[23] The world is an ocean of poison, a sea of illusion, across which only the wise guru can navigate and guide the devotee. If true belief is *amrita*, then poison, 'the liquor of folly', is its malign antithesis.[24] These sacred verses display a common religious tendency to associate women, and the lust they arouse in men, with moral and spiritual poison. But several of them also depart from a purely metaphorical idiom by identifying poison with specific plants, seeds and fruits, thereby conflating poison as moral message with poison as material object.[25]

Looking forward in time for a moment, it is hardly surprising that the mythic modelling and scriptural appropriation of poison found its way into later literary convention and the political imagery of the nineteenth and twentieth centuries. In 1873, the Bengali writer Bankim Chandra Chatterjee published a novel entitled *Bisha Briksha* ('The Poison Tree'), which drew upon the popular idea of a poison tree that (like the poison maiden) fatally infected and corrupted all those who came near it: this was the 'upas tree' of European literature and legend.[26] In Bankim's novel, the poison tree symbolizes human desire, anger and envy. In the words of the author, 'The want of self-control is the germ of the poison tree, and also the cause of its growth. This tree is very vigorous; once nourished, it cannot be destroyed. Its appearance is very pleasing to the eye … But its

---

[21] Siegal, *Sacred*, 149, 27, 274.

[22] Ernest Trumpp (ed.), *The Adi Granth, or the Holy Scriptures of the Sikhs* (London: William Allen, 1877), 53, 59, 106, 129.

[23] Ibid., 250, 276.   [24] Ibid., 499, 561.   [25] Ibid., 206.

[26] On the 'upas' myth, see Henry Yule and A. C. Burnell, *Hobson-Jobson: A Glossary of Colloquial Anglo-India Words and Phrases* (2nd ed., London: Routledge & Kegan Paul, 1985), 953–59. The real upas tree (*Antiaris toxicaria*) grew in India's Western Ghats and had medicinal as well as toxic properties: R. N. Chopra, R. L. Badhwar and S. Ghosh, *Poisonous Plants of India* (New Delhi: Indian Council of Agricultural Research, 1965), 2: 811–12.

fruit is poisonous; who eats it dies.'[27] In the story, Nagendra, a wealthy young landlord (zamindar), becomes smitten with the young orphan girl Kunda whom he has adopted, causing his devout, self-sacrificing wife, Surya Mukhi, to try to escape her emotional turmoil by leaving home. Nagendra belatedly realizes his mistake, and Surya returns home from her ordeal of separation and wandering to find a repentant husband. Bankim ends the story remarking: 'The "Poison Tree" is finished. We trust it will yield nectar in many a house.'[28] It gives added weight to this moral tale that Bankim took a keen personal interest in Western medical ideas and their application to everyday life. In 1878, he published in *Bangadarshan*, the Bengali journal he had founded six years earlier, an account of a 'toxicological chart' compiled by a Calcutta Medical College graduate, Harishchandra Sharma. Intended for hanging on the wall for household use, the chart gave a brief description of various metallic, herbal and animal poisons, their symptoms and treatment.[29] In the 1870s, poison (as both a material and a moral entity) and its potential dangers within the home were evidently much on Bankim's mind.

The rhetorical invocation of poison did not end with the nineteenth century. Mohandas Gandhi frequently used the image of poison not just to formulate a moral agenda like Bankim did, but also to make his anti-colonial message more compelling. In 1909, he likened sexual vice and the passion for money to a poison, one that was worse than the bite of a snake: snake venom 'merely destroys the body' whereas lust and greed 'destroy body, mind and soul'.[30] In a speech made in 1916, following his return from South Africa, he remarked: 'England has sinned against India by forcing free trade upon her. It may have been food for her, but it has been poison for this country.'[31] When violence erupted at Chauri Chaura in northern India in February 1922, and threatened to engulf Gandhi's non-violent civil disobedience movement, the Mahatma inveighed against 'the crime', and then 'the poison', of Chauri Chaura.[32] Two decades later, speaking at Patna in 1947, he argued that while foreign mill cloth was 'like poison', handmade homespun cloth (*khadi*) was 'like

---

[27] Bankim Chandra Chatterjee, *The Poison Tree: A Tale of Hindu Life in Bengal* (London: T. Fisher Unwin, 1884), 191–92.

[28] Ibid., 314.

[29] *Bangadarshan* 6 (1285 Bengali era): 101–03, with thanks to Projit Bihari Mukharji.

[30] M. K. Gandhi, *Hind Swaraj* (1909) in Anthony J. Parel (ed.), *M. K. Gandhi: Hind Swaraj and Other Writings* (Cambridge: Cambridge University Press, 1997), 108.

[31] Speech at Madras, 14 February 1916, [M. K. Gandhi], *Collected Works of Mahatma Gandhi* 13 (New Delhi: Publications Division, Ministry of Information and Broadcasting, 1964), 223.

[32] [M. K. Gandhi], *Gandhi's Speeches and Writings* (Madras: G. A. Natesan, n.d.), 657.

nectar'.[33] A device to differentiate the Indian self from its colonial other, poison also had its post-colonial usages. Shortly after Gandhi's assassination in January 1948, India's prime minister, Jawaharlal Nehru, made an impassioned speech in Delhi, calling on India to rid itself of the 'poison' of communalism. 'The flow of poison', he warned, 'if not checked immediately, was sure to lead the country to even greater disasters'.[34]

Poison and the healing art

The Sanskrit word *visha* – the *bish* of north Indian vernaculars and springing from the same etymological root as the English word 'vicious' – could signify all that was evil, destructive and corrupting. But a notion of *bish* as both poison and cure was widely present in Indian medicine, from the written texts of Ayurveda and Unani medicine through to the many variants of folk medical practice and belief.[35] Indeed, India's medical traditions can be categorized as being as much systems of toxicology – or poison management – as of therapeutics. As stated in the *Caraka Samhita*, 'Even acute poison is converted into an excellent medicine by the right method of preparation; while even a good medicine may act as an acute poison if improperly administered'.[36] Or, as paraphrased by Udoy Chand Dutt, 'Taken in large doses, poisons destroy life, but, judiciously used, they act as curatives and restore health, even in dangerous diseases'.[37] Toxicity might thus be a power worthy of embrace, not a poison damsel to be shunned.

Poisons figured prominently in early Ayurvedic texts, their properties, symptoms and antidotes forming one of the eight principal branches of medical knowledge. Great importance attached to the *vaid* (physician) having an extensive knowledge of poisons so as to serve the king for whom poisoning was 'a peril from which he is rarely free'.[38] Ayurveda divided *visha* into several categories. These included the broad characterization of

---

[33] Address to Workers, Patna, 24 April 1947, [M. K. Gandhi], *Collected Works of Mahatma Gandhi* 87 (New Delhi: Publications Division, Ministry of Information and Broadcasting, 1983), 349.

[34] *ToI*, 3 February 1948, 8.

[35] The idea of medical systems in India has been much criticized, partly on the grounds that medical beliefs and practice exhibited enormous internal differences and regional variations, exchanged ideas and practices among themselves (and latterly with Europe) and were only coherent 'systems' in Western eyes. The concept remains useful, however, in addressing the broad divisions in medical belief, practice and agency in India.

[36] *Caraka Samhita* (6 vols, Jamnagar: Shree Gulabkunverba Ayurvedic Society, 1949), 2: 27.

[37] Udoy Chand Dutt, *The Materia Medica of the Hindus* (Calcutta: Thacker, Spink, 1877), 7.

[38] Henry R. Zimmer, *Hindu Medicine* (Baltimore: Johns Hopkins University Press, 1948), 85–86.

a poison according to its plant, mineral or animal origin, further differentiated by its source within a given plant (such as roots, leaves, fruits and tubers) or from a particular animal part or product (saliva, faeces, vomit, menstrual fluid). Poisons were variously said to be rough, hot, quick, penetrating, slow and subtle. As therapeutic substances, their nature and functions were classed according to their physical effects on the human body, understood within a humoral system of physiology and diagnostics in terms of wind, bile and phlegm, while in their toxic guise they were thought of as attacking and vitiating the essential organs and tissues of the body and so extinguishing the elements that gave or sustained life.[39] In Ayurveda and across Indian medicine more generally, poisons were thought of as substances that had a 'heating' or stimulating effect. Their 'hot' energy excited the body, enabling it to overcome lethargy, impotence and frigidity.[40] Given in therapeutic doses, such substances were not inherently toxic: they acted as stimulating tonics rather than life-threatening drugs. Indeed, their potency was valued for the added efficacy they brought to healing medicaments and compounds designed to counter dangerous or intractable complaints, as in the use of arsenic and mercury to treat leprosy.[41]

Turning a lethal substance into a medicinal one was an art that required the knowledge and skill of a trained physician – or at least one well versed in the texts. In many drug preparations, the technique lay in knowing ways to manage toxicity and turn its dangerous potency to sound therapeutic use. This might be done by combining the raw drug with other substances so as to harness its strength while moderating its toxicity, or by marinating, soaking and 'cooking' the ingredients so as to render them fit for human use. In 1826, in one of the first European descriptions of the preparation of Indian medicines, H. H. Wilson detailed the treatment of cholera patients by Bengali *vaids*. According to his Indian informant, if the disease failed to respond to initial medical intervention, the physician, having first sought the permission of relatives, would turn to animal or vegetable poisons. The ingredients in these potions included cobra venom and *bish* or *bisk* (here meaning the drug aconite from the roots of *Aconitum ferox*), made up into pills with such formidable names as the 'death-destroying pill' and 'the recovery of the dead'. One of the most popular medicaments was a compound prepared from *bish*, red and yellow arsenic, mercury, mica, sulphur and vermilion, steeped in lime juice, ginger and cannabis. The mixture was boiled, cooled, beaten into a

[39] Sundar, *Treatment for Poisons*, 5–8.
[40] Francis Zimmermann, *The Jungle and the Aroma of Meats: An Ecological Theme in Hindu Medicine* (Berkeley: University of California Press, 1987), 112, 122.
[41] H. H. Wilson, '*Kushta*, or Leprosy, as Known to the Hindus', *TMPSC* 1 (1825): 43–44.

paste and combined with animal and fish gall. From this, tiny pills were prepared (small enough to pass through the eye of a needle) and administered with a cooling draught of coconut milk. The medicine was said to raise the pulse rate and encourage natural heat to return to the body (which in cholera became deadly cold). If necessary, a second dose was administered.[42] In 1877, fifty years after Wilson, Udoy Chand Dutt placed a chapter on poisons at the very start of his account of the 'materia medica of the Hindus', as if to demonstrate the unapologetic centrality of toxicological knowledge to indigenous medical practice. Like Wilson, Dutt gave a detailed description of how medicines containing poisons were prepared, including one which required a portion of *Aconitum* root to be purified by steeping in cow's urine for three days before use. Another preparation called for aconite to be mixed with sulphur, black pepper, borax and cinnabar, before being made up into pills as a febrifuge.[43]

How far such elaborate instructions and intricate techniques were followed in practice is unclear. But such accounts do show that the therapeutic, as well as toxic, properties of metals and minerals were well known to pre-colonial India. However, as many of these substances were not used or available in a pure form, understanding of their toxic potency was perhaps limited. Mercury was principally used in the form of cinnabar (sulphide of mercury). Arsenic was most widely known through its sulphides, red arsenic (realgar) and, more especially, yellow arsenic (orpiment), which were far less toxic than pure white arsenic. It further appears that a number of metallic poisons like mercury, as well as vegetable drugs like opium, were absent from, or little used in, ancient Ayurveda. They became more common with the arrival of Muslim hakims, practitioners of the Unani ('Greek' or Hippocratic) system of medicine, from the thirteenth century onwards.[44] In most Ayurvedic texts, vegetable and animal poisons far outweighed minerals in their number and utility. But India's toxic knowledge did not flourish in isolation. Poison-lore circulated for centuries between India and the Middle East: several Indian treatises on poison and other branches of medicine, some now lost in the original, were translated into Persian and Arabic between the eighth and twelfth centuries. Among these was an influential text attributed to an Indian author identified only as 'Shanaq': his work in turn formed a major source for Ibn Wahshiya's *Book on Poisons* in ninth-century Iraq.[45] Long before the *Arabian Nights*

[42] H. H. Wilson, 'On the Native Practice in Cholera', *TMPSC* 2 (1826): 284–87.
[43] Dutt, *Materia Medica*, 96–97.   [44] Ibid., xi, 23.
[45] Martin Levey, 'Medieval Arabic Toxicology: The *Book of Poisons* of Ibn Wahshiya and Its Relation to Early Indian and Greek Texts', *Transactions of the American Philosophical Society* 56 (1966): 6–10.

reached Europe and infiltrated the Western imagination, Middle Eastern texts had already assigned an exceptional potency to Indian poisons. Alongside 'poison maidens', they noted such powerful drugs as 'Indian aconite' (*bish*), and poison mixtures of Indian origin such as *bishrahi*.[46] Indian works were further cited by Middle Eastern writers as authorities on 'sex potions' and aphrodisiacs.[47]

Medical texts and treatises, whether by Muslim or Hindu authors, continued into early modern times to make reference to specific poisons, their uses, symptoms and antidotes, though the extent of their description and the importance ascribed to particular substances varied widely. A seventeenth-century text by Noureddeen Mohammed Abdullah Shirazi, court physician to the Mughal emperor Shah Jahan, made only passing reference to wolf's bane (a drug of the aconite family) and other poisons.[48] A century later, another Unani treatise, translated into English as the *Taleef Shereef*, referred to a number of vegetable, mineral and animal substances that were identified as having both therapeutic and toxic properties, including aconite, datura, orpiment and nux vomica – the latter the source of deadly strychnine.[49]

Beyond the formal texts and orthodox practices of Ayurvedic and Unani practitioners, there existed a vast popular poison-lore and a great range of practitioners from rural *vaid*s and hakims through itinerant sadhus and fakirs to individual villagers – women and men – who were renowned for their healing skills. For instance, leaves of the datura plant, a drug that features prominently in the following chapters for its criminal associations, were pounded and mixed with turmeric to make a cooling paste applied to inflamed parts of the body. The leaves were mixed with opium and oil to remove body lice and cure skin diseases, or made up into pills to treat toothache. Datura leaves were smoked to relieve asthma and other respiratory ailments.[50] Given the plant's wide distribution, such medicinal uses were common, did not require formal doctoring, and belied the notoriety that became attached to datura in colonial times.

[46] Ibid., 14–15, 85, 118.
[47] David L. Newman (ed.), *The Sultan's Sex Potions: Arab Aphrodisiacs in the Middle Ages* (London: Saqi Books, 2014), 33–35.
[48] Francis Gladwin, *Ulfaz Udwiyeh, or the Materia Medica in the Arabic, Persian, and Hidevy Languages Compiled by Noureddeen Mohammed Abdullah Shirazy* (Calcutta: Chronicle Press, 1793).
[49] George Playfair, *The Taleef Shereef, or Indian Materia Medica* (Calcutta: Medical and Physical Society, 1833), 29–31, 81–82, 107–08.
[50] William Dymock, *Pharmacographia Indica: A History of the Principal Drugs of Vegetable Origin, Met With in British India* (3 vols, Calcutta: Thacker, Spink, 1890–91), 2: 586.

The social function of poison

Within the wide ambit of pre-colonial medicine, toxicology was an established art and poison management a significant mode of therapeutic activity. In other social and cultural realms, toxic substances had a seemingly less prominent role, but this did not make them unimportant. Indeed, to think about the social uses of poison in pre-British and early colonial India is to reflect on the extent to which a knowledge of poisons was widely dispersed across society, as a matter of popular, and not merely elite, understanding and practice. But it is often difficult to document the nature and extent of this social engagement with poison except through colonial sources that were hostile to its use. Poisons and intoxicants of one description or another came to be implicated in what J. C. Marshman referred to as those 'barbarous customs', 'atrocious rites' and 'criminal acts' – from female infanticide through sati to *thugi* – which Western commentators so decried and saw as an urgent rationale for colonialism's 'mission of humanity', and whose extirpation was presented as proof of Europe's 'benevolent labours'.[51] Precisely because of the way in which commonplace drugs like opium and datura were drawn into this condemnatory rhetoric and the science it spawned, it is difficult to reconstruct their everyday use except through sources colonial rule itself generated.[52] Any attempt to present the social life of substances is thus tainted by colonialism's epistemological quest and politicizing agenda, as poisons and poisoning became subject to the regime's ideology, policing and judicial and forensic processes.

If we take, for instance, what colonial sources described as the 'peculiar and unnatural crime' of female infanticide, then opium poisoning was identified as one of the means by which unwanted girl children met their deaths in the Rajput lineages of western, central and northern India, where that practice was followed.[53] Contemporary accounts include reference to asphyxiation, drowning, starvation and general neglect, but

---

[51] John Clark Marshman, *The History of India* (3 vols, London: Longmans, Green, Reader & Dyer, 1867) 3: 51, 59, 104, 107.

[52] Daniel J. R. Grey, 'Creating the "Problem Hindu": *Sati*, Thuggee and Female Infanticide in India, 1800-60', *Gender and History* 25 (2013): 498–510.

[53] Charles Raikes, *Notes on the North-Western Provinces of India* (London: Chapman & Hall, 1852), 4. On female infanticide, see Malavika Kasturi, 'Law and Crime in India: British Policy and the Female Infanticide Act of 1870', *Indian Journal of Gender Studies* 1 (1994): 169–93; idem, 'Taming the "Dangerous" Rajput: Family, Marriage and Female Infanticide in Nineteenth-Century Colonial North India', in Harald Fischer-Tiné and Michael Mann (eds), *Colonialism as Civilizing Mission: Cultural Ideology in British India* (London: Anthem Press, 2004), 117–40; Satadru Sen, 'The Savage Family: Colonialism and Female Infanticide in Nineteenth-Century India', *Journal of Women's History* 14 (2002): 53–79.

opium was reported as one of the main methods used to extinguish infant lives. It is not hard to find reasons for this. Opium was a common household commodity across the region where female infanticide occurred. It was used among Rajputs as a narcotic and domestic remedy for various ailments including fever: indeed, there were few locally available substances that had a wider range of medical applications. Its capacity to kill (especially when given to infants) was clearly not the reason for its domestic presence, just as opium and laudanum were present in Victorian households in Britain without being thereby intended for suicide and murder. Opium was representative of the way in which in India, as elsewhere, poisons were everyday substances put to exceptional uses. Unlike some of India's more ferocious drugs, such as aconite and nux vomica, opium promised a milder manner of death, one more akin to sleep than murder. Moreover, while within a patriarchal social system it was the male heads of Rajput households who directed or approved the killing of infant girls – for reasons of caste status and social prestige rather than marry them into families socially inferior to their own – they did not do the killing themselves. That was a task delegated to women. As an Indian informant told the British political agent in Gujarat in 1808, it was an 'affair of the women' and 'no part of the business of men'. Opium was smeared on the mother's nipples or kneaded into a small ball and inserted into the baby's mouth by *dais* (midwives) and other household servants with ready access to opium.[54]

The identification of poisoning with women, rather than men, can be seen as one of its most striking trans-cultural characteristics, though how far this represents a negative stereotype of women or reflects a social reality in which women were more frequently poisoned than men is open to question.[55] According to one textbook of medical jurisprudence, women in India, 'more so than in Europe, employ poison rather than bodily violence, and their crime is directed for the most part against their husband, or some rival in his affections'.[56] Certainly, to omit poison from the social history of India would be to deny women (and men) a kind of

---

[54] Alexander Walker to Jonathan Duncan, Governor of Bombay, 15 March 1808, in Edward Moor, *Hindu Infanticide: An Account of the Measures Adopted for Suppressing the Practice of the Systematic Murder by Their Parents of Female Infants* (London: J. Johnson, 1811), 53.

[55] Ian Burney, *Poison, Detection, and the Victorian Imagination* (Manchester: Manchester University Press, 2006), 21–24. Katherine Watson, *Poisoned Lives: English Poisoners and Their Victims* (London: Hambledon Continuum, 2004), 45, notes that 'contrary to popular opinion' in Victorian England, the number of male and female poisoners was roughly equal. On women poisoners in Britain, see George Robb, 'Circe in Crinoline: Domestic Poisonings in Victorian England', *Journal of Family History* 22 (1997): 176–90.

[56] Waddell, *Lyon's*, 29–30.

agency in their own lives as well as to overlook a significant marker of their subordination and victimhood. In India, men were rarely reported as having poisoned women. This is not to suggest that uxoricide did not happen: it clearly did, but, with other forms of violence available to men, many of them socially sanctioned, arsenic and opium were not the usual means of doing so.

It was observed in the Introduction that statistical evidence for poisoning during the colonial era was often scanty. Female infanticide is a case in point. If the practice was as widespread as many British commentators believed, then opium poison was implicated in the death of thousands (even tens of thousands) of female children every year and thus a significant factor in depressing India's population overall and in creating a marked gender imbalance in which, across large swathes of northern and western India, males greatly outnumbered females.[57] Of course, it was infanticide as such, rather than the instrumental use of opium, that the British sought to suppress and which became the subject of the 1870 Female Infanticide Act, and yet by association drugs like opium and datura came to be seen as the embodiment of a singularly Indian barbarity and a specific danger to human life.

Suicide, like murder, can reasonably be assumed to have had many different motives, but most of these passed unrecorded in India's colonial archive. According to Robert Harvey, 'One man killed himself because he was blind and helpless; another on account of a disagreement about some land; a third on being arrested on a charge of theft; a fourth because he was out of work and tired of life; but the motives, as a rule are not given'.[58] Although Harvey referred only to men's motives, it was more usually women than men who took their own lives. They might find any number of means to do so. Fanny Parks, living in Allahabad in 1828, was concerned about the oleanders growing in her garden, fearing that horses and cows might eat their poisonous leaves, but more especially that they would be used by Hindu women, who, 'when tormented by jealousy, have recourse to this poison for their self-destruction'.[59] But opium was

---

[57] Walker's informant suggested that 20,000–30,000 female Rajput infants perished in Gujarat in this way every year, but admitted that there were no firm data: Moor, *Hindu Infanticide*, 60.

[58] Robert Harvey, 'Report on the Medico-Legal Returns Received from the Civil Surgeons in the Bengal Presidency during the Years 1870, 1871, and 1872', *IMG* 11 (1876): 60. On suicide, see also W. J. Buchanan, 'A Chapter on Medical Jurisprudence in India', in Fred. J. Smith (ed.), *Taylor's Principles and Practice of Medical Jurisprudence* (7th ed., London: J. & A. Churchill, 1920, 2 vols), vol. 2, 894–95, 894–95; T. E. B. Brown, *Punjab Poisons* (3rd ed., Lahore: 'Civil and Military Gazette' Press, 1888), 119–23.

[59] Fanny Parks, *Wanderings of a Pilgrim in Search of the Picturesque* (2 vols, London: Pelham Richardson, 1850), 1: 78.

most commonly the substance employed. In an analysis of suicides in one district of the North-Western Provinces in 1891, G. D. McReddie reported that just over half of recent victims (97 out of 180) had taken opium and nearly 80 per cent of these were women. He found their use of opium easy to explain: opium poppies grew in the fields, and opium was a common household commodity. 'Hanging and drowning require a certain amount of forethought and preparation', he added, while opium was 'a means close to hand; all that is required is to swallow a small quantity and the passage to another world is easy and speedy'.[60]

Like female infanticide, suicide by opium showed women as both perpetrators and victims but within a patriarchal system that assigned women low status and allowed them few independent choices. One might set against this the many reported cases in which a young wife, sometimes at a lover's instigation, tried to kill her much older husband by putting arsenic in his food and, in this limited sense, might be said to have defied patriarchy.[61] This suggests the possibility of viewing poison as, actively or implicitly, an act of 'everyday resistance'.[62] But all too commonly, the consequence of oppression and unhappiness was suicide, an act that turned the agency of poison not against father, husband and in-laws but against the lonely, shamed, troubled woman herself.

Unlike some of the poison crimes considered later in this book, one can sometimes detect a degree of sympathy for women who resorted to suicide, abortion and murder. In the colonial system of justice, empathy seldom ensured exemption from punishment, but it might favour its modification. For instance, in 1830, a 16-year-old woman named Fatima was charged with trying to murder her husband with arsenic and lead acetate: the appeal judges in Bombay reduced the lower court's sentence of ten years' imprisonment to 18 months on account of her youth and the fact that her husband had survived her attempts to poison him.[63] In another Bombay case two years earlier, Johre Kome Babnya was accused of murdering her husband 'by putting poison into his *kitcheree* [kedgeree], of which he partook and died the same evening'. When it was claimed that she wanted to kill her husband in order to live with her lover, she countered by alleging that she was being falsely accused because she

---

[60] G. D. McReddie, 'Opium Suicides in Hardoi District', *IMG* 26 (1891): 168.

[61] For one such case, involving a 15-year-old Muslim girl and the death by arsenic poisoning of her older husband, see *ARCED (Bengal), 1915*, 6.

[62] James C. Scott, *Weapons of the Weak: Everyday Forms of Peasant Resistance* (New Haven, CT: Yale University Press, 1985). Cf. Giles, 'Poisoners', 108: 'Poison is naturally the weapon of the weak.'

[63] For domestic poisonings, see A. F. Bellasis (comp.), *Reports of Criminal Cases Determined in the Court of Sudder Foujdarree Adawlut of Bombay* (Bombay: Government Press, 1849), 37–38.

had refused her brother-in-law's 'incestuous solicitations'. The sessions judge sentenced her to two years' imprisonment, but the judges of the superior court questioned the evidence in the case, especially the absence of a post-mortem, believed Babnya's 'confession' had been forced out of her, and quashed her conviction.[64]

The use of poisons – or, more exactly, what came in the course of the nineteenth century to be designated poisons – was often deeply paradoxical. Substances like opium, aconite and even arsenic were widely regarded as having aphrodisiac powers. By the late nineteenth century, 'love potions' of this kind were not only prescribed by local *vaids* and hakims but also widely advertised – and, in British eyes, scandalously displayed – for sale in the Indian press.[65] Such substances might be sought after as an aid to sexual prowess and pleasure but more commonly they were seen as medicaments, a cure for impotence and infertility (but also for syphilis) and as an aid to procreation, and hence the fulfilment of the social imperative to bear children, especially male ones. Reflecting a staunchly masculine view of the world and the gendered function of therapeutics, the *Caraka Samhita* left readers in no doubt that the purpose of aphrodisiacs was to stimulate male sexual desire, increase the flow of semen and so create progeny.[66] The paradox is that what functioned as a poison – as a life-destroying force – in one context served in another as an empowering substance – to facilitate the creation of life itself. To extend the paradox still further, the aphrodisiacs and elixirs in popular use were identified by practitioners of Western medicine as a source not of pleasure but, by dint of their toxic ingredients, of death or serious injury by poisoning. And the same substances that were touted as being conducive to pleasure and procreation in one context were in another among those used, in suicide, infanticide and husband murder, to end life. In this topsy-turvy world, wives who murdered their spouses with arsenic or aconite could claim in their defence that they had intended the 'medicine' that killed their husband as a 'love philtre' to regain his waning passion or to ensure the birth of a much-needed child.[67]

The observation about poison's ambiguous nature and paradoxical effects can be taken a stage further if we turn to abortions. Like female infanticide, the use of poisons to cause abortion can be seen not only as a

---

[64] Ibid., 16–18.
[65] Charu Gupta, *Sexuality, Obscenity, Community: Women, Muslims, and the Hindu Public in Colonial India* (Delhi: Permanent Black, 2001), 72–73, 80; Deana Heath, *Purifying Empire: Obscenity and the Politics of Moral Regulation in Britain, India and Australia* (Cambridge: Cambridge University Press, 2010), 175–76, 197.
[66] *Caraka Samhita* (5 vols, Delhi: Sri Satguru Publications, 1996), 3: 700.
[67] C. H. Bedford, 'Notes of Some Toxicological Experiences in Bengal and in the Punjab', *IMG* 37 (1902): 204.

further example of a toxicity operating within a female domain of knowledge and practice, but also as simultaneously serving to uphold patriarchal social values and caste authority. Abortions were largely sought after by women to destroy the foetal evidence of proscribed or involuntary sexual relations – between members of the same family (brothers-in-law, fathers-in-law), between men and women of different castes (where marriage between castes was proscribed) or between widows and their lovers in a society where women married at a very young age but on becoming widows were prohibited from remarrying. In the view of one colonial medical officer discussing abortion, 'The prohibition of widow-marriage is ... at the bottom of these evils', adding that 'fear of excommunication [from their caste] leads the unfortunate creatures to become either active or passive agents in the crime'.[68] Another doctor remarked that 'The system of infant marriages and enforced widowhood have a great deal to do with the frequent recourse to abortive agents'.[69]

The extent to which abortion was practised to prevent the birth of an unwanted child is difficult to establish for any society, especially before legislation was introduced to legalize its use. Abortion was illegal under colonial law and regarded as a 'defining moral sin', on a par with killing a Brahmin, in Hindu religious texts.[70] In 1854, C. R. Baynes expressed a widely held view when he remarked that 'It may be feared that this crime [abortion] prevails in this country to a far greater extent than we have any accurate idea of'.[71] In 1920, W. J. Buchanan similarly observed: 'It is impossible to obtain statistics of the degree of prevalence of this offence, as it is only the fatal cases that come to the notice of the police.'[72] The bodies of women who died following an abortion were often concealed or their deaths attributed to other causes. Among the 361 medico-legal cases referred to Bengal's chemical examiner in 1873–74, 45 involved suspected abortion.[73] In the 1880s, Dulip Singh, on leave in rural Punjab, heard of fifteen abortions being carried out in two adjacent villages (with a combined population of around 4,000) over a two-month period – all performed by one old woman.[74] Abortionists were mostly village women, especially *dais* – low-caste midwives – who were themselves an object of

[68] V. Richards, 'Criminal Abortion', *IMG* 6 (1871): 230.
[69] Dulip Singh, 'Modes of Inducing Criminal Abortion in the Punjab', *IMG* 20 (1885): 9.
[70] Doniger, *Hindus*, 572.
[71] C. R. Baynes, *Hints on Medical Jurisprudence, Adapted and Intended for the Use of Those Engaged in Judicial and Magisterial Duties in British India* (Madras: Pharoah, 1854), 128.
[72] Buchanan, 'Medical Jurisprudence', 896.
[73] Supriya Guha, 'The Unwanted Pregnancy in Colonial Bengal', *Indian Economic and Social History Review* 33 (1996): 412–13. On the difficulty of detecting abortion, see Harvey, 'Report', 145.
[74] Singh, 'Modes', 9.

intense colonial hostility for their crude, violent and unhygienic birthing techniques. Their association with abortion further diminished their standing among colonial medical officers.[75] But bazaar apothecaries and druggists (*pansaris*) also supplied the necessary drugs where these were not found growing locally.

Various methods were used to cause an abortion. Some sources suggested that abortion was 'largely done by drugs'; others questioned this, suggesting that external, 'mechanical' means were more common.[76] 'Mechanical' methods involved putting pressure on a woman's abdomen, or forcing twigs or a wodge of cotton and other objects into her uterus. Ingestion of, or internal exposure to, toxic substances, presented in the guise of 'medicines', was another technique, and the materia medica of British India contain a large number of substances, most of vegetable origin, used for this purpose. The root of *lal chitra* (*Plumbago rosea*) was the drug most commonly cited, but a host of other plant substances such as oleander and marking nut were used, as were quicklime, arsenic and potassium carbonate.[77] These toxic doses brought on violent spasms, retching, vomiting and purging. Smeared on suppositories and thrust into the uterus, poisons caused severe internal inflammation and violent uterine contractions. Given the crude manner in which they were performed, abortions were life threatening – alike to mother and foetus. Women suffered severe internal injuries (particularly from perforation of the wall of the uterus) and died an intensely painful death from blood poisoning and peritonitis.[78] Nowhere, perhaps, than in abortion was the violence of poison made more apparent.

[75] On *dai*s, see Geraldine Forbes, 'Managing Midwifery in India', in Dagmar Engels and Shula Marks (eds), *Contesting Colonial Hegemony: State and Society in Africa and India* (London: Academic Press, 1994), 152–72; Sean Lang, '"Drop the Demon *Dai*", Maternal Mortality and the State in Colonial Madras, 1840–1875', *Social History of Medicine* 18 (2005): 357–78.

[76] Brown, *Punjab Poisons*, 5; [Anon.], 'The Poisons Used to Destroy Human Life in Bengal', *IMG* 20 (1885): 321.

[77] Kanny Lall Dey, 'Medicinal Substances Used by Native Practitioners', in A. M. Dowleans, *Official Classified and Descriptive Catalogue of the Contributions from India to the London Exhibition of 1862* (Calcutta: Bengal Printing Co., 1862), 75–77; *RCA (Bombay), 1874–75*, 11. On *lal chitra*, see J. C. Lisboa, 'Famine Plants: Wild Herbs, Tubers, Etc. Used as Food during Seasons of Scarcity', in *Gazetteer of the Bombay Presidency. Vol. 24: Botany* (Bombay: Government Central Press, 1886), 266; John D. Gimlette, *Malay Poisons and Charm Cures* (3rd ed., Kuala Lumpur: Oxford University Press, 1971), 201–04.

[78] Richards, 'Criminal Abortion', 230–31; Ranajit Guha, 'Chandra's Death', in Partha Chatterjee (ed.), *Ranajit Guha, The Small Voice of History: Collected Essays* (Ranikhet: Permanent Black, 2009), 271–303; Indira Chowdhury, 'Delivering the "Murdered Child": Infanticide, Abortion, and Contraception in Colonial India', in Deepak Kumar and Raj Sekhar Basu (eds), *Medical Encounters in British India* (New Delhi: Oxford University Press, 2013), 275–98.

Poverty and poison

Poison had as close, and as troubling, a relationship with poverty as it did with gender. In the course of the nineteenth century, India was convulsed by a series of major famines. Few areas of the country were immune to hunger, dearth and the epidemics of smallpox, cholera, malaria and dysentery that accompanied famine or crowded in its wake. Millions died from starvation, disease and their combined effects. A common factor in India's famines was the failure of food crops or a fatal combination of food scarcity and high prices with the loss of employment and purchasing power. One of the many responses of famine-struck populations was thus to look beyond their normal means of subsistence, beyond field crops and the marketplace, and to forage instead for 'famine foods'. These 'surrogate' foods might be found growing on the margins of cultivation, on waste ground, in forests and jungles. Some were wild grasses, grains and greens, while others were the roots, fruits and leaves of plants that were familiar to consumers in normal times, but used only sparingly or seasonally to supplement more wholesome foods.[79] But famine necessitated the eating even of ill-favoured foods normally considered too bitter or irritant for use. Others required cooking to make them palatable – a task that was impossible in desperate times when fuel was in short supply, cooking utensils had been sold and famine-struck survivors were too weak to do more than gather and eat what they found in the wild. Some plants were actively toxic: one account of famine in mid-1890s Bengal refers to tribal Santhals emerging from the forest with armfuls of wild plants, 'enough', an observer wrote, 'to poison a regiment'.[80] Some surrogate foods, even if they did not kill outright, caused severe diarrhoea or acute irritation to the gut and bowels, further weakening the hungry and malnourished. As with the use of poison in infanticide and abortion, the extent of the debility and mortality caused by recourse to famine foods is impossible to quantify but it is bound to have increased the already soaring number of deaths from starvation and disease.

Most contemporary accounts of famine in India make some reference to the collecting and eating of these emergency foods. Reporting on their nature, use and physiological effects became, by the second half of the nineteenth century, a task taken up by many colonial physicians and

---

[79] A. H. Church, 'Vichka Seed as a Famine Food in the Bombay Presidency', *Agricultural Ledger* 6 (1899): 1–2; O. Reinherz, 'The Seeds of *Shorea robusta* as a Famine Food', *Agricultural Ledger* 11 (1904): 33–36.

[80] Malabika Chakrabarti, *The Famine of 1896–1897 in Bengal: Availability or Entitlement Crisis?* (Hyderabad: Orient Longman, 2004), 319.

botanists and their Indian counterparts.[81] In 1886, José Camillo Lisboa, a Goan botanist resident in Bombay and already the author of a pioneering study of the 'useful plants' of the province, published a detailed account of 'famine plants . . . used as food during seasons of scarcity'. His evidence came principally from the famine of 1876–78, during which, he noted, many British officials had for the first time become aware of the extensive range of 'wild herbs' resorted to by the poor 'for want of ordinary food'. Lisboa speculated that, while unfamiliar to Europeans and even to most city-dwelling Indians, knowledge of these plants – he listed nearly a hundred – had perhaps been the 'result of [the] accumulated experience of bygone generations', exposed to repeated episodes of drought, conquest and failed crops.[82] In 1906, Chunilal Bose, the Chemical Examiner in Calcutta, investigated the toxic principles of one such famine food, the fruit of the *dhoondool* plant (*Luffa aegyptiaca*). This, he reported, was occasionally eaten by poor people in Bengal, after being repeatedly washed and boiled to remove its bitter taste and poisonous content. Cultivated varieties of the plant tasted sweet and were relatively harmless, but the uncultivated form, gathered as a famine food, was bitter and poisonous and caused severe vomiting and diarrhoea.[83]

Twenty years divided Lisboa's descriptive list from Bose's more scientifically exacting analysis of *dhoondool* and its toxic properties: in that period botanical and biochemical enquiry had made considerable advances. But it is significant that in both cases Indians – a Goan and a Bengali – were motivated to undertake scientific enquiries into the foods of the poor and chose to position themselves between Western science (and European ignorance) on the one hand and 'native' knowledge and experience on the other. At the same time, such studies showed how, by investigating nutrition and poisoning among the poor, colonial medicine (and colonial science more generally) was able to move out of the enclave of European need and European agency to which it had hitherto been largely confined.[84] Poverty, in its more literally toxic manifestations, was on its way to becoming a subject of scientific scrutiny and informing a wider discourse of deprivation, health and nutrition.

---

[81] David Arnold, 'Famine in Peasant Consciousness and Peasant Action: Madras 1876-8', in Ranajit Guha (ed.), *Subaltern Studies III* (Delhi: Oxford University Press, 1984), 94–95.

[82] Lisboa, 'Famine Plants', 190–91.

[83] Chunilal Bose, 'The Toxic Principles of the Fruit of *Luffa aegyptiaca*', in J. P. Bose (ed.), *The Scientific and Other Papers of Rai Chunilal Bose Bahadur* (Calcutta: Forward Press, 1924) 1: 86–103.

[84] On enclavism, see Arnold, *Colonizing*, ch. 2. For nutritional science, see David Arnold, 'The "Discovery" of Malnutrition and Diet in Colonial India', *Indian Economic and Social History Review* 31 (1994): 1–26.

It was a short step from emergency foods, resorted to in times of dearth and famine, to foodstuffs that were grown and consumed on a more regular basis but whose poisonous nature was known to their consumers. The principal example of this was lathyrism, the condition caused by consumption of a pulse known as *kesari*, from the vetch *Lathyrus sativus*, which contains a powerful neurotoxin. This plant was grown – or occurred in a semi-wild state – in many parts of central and eastern India, especially on poor soils where no other crops would thrive. In the form of dal, it was consumed by extremely poor peasants, whose livelihood was dependent upon landlords who controlled their labour and kept them in abject poverty. *Kesari* was one of the few means by which labourers were rewarded for their toil or the only crop that would grow, especially in rain-deficient seasons, on the meagre land assigned to them. The consequence of such extreme reliance upon a single food source was a form of paralysis that caused progressive loss of mobility in the victims' lower limbs and a slow death.

References to lathyrism in Sanskrit medical texts can be dated back to the sixteenth century.[85] The first British observations were made by the East India Company surgeon-naturalist Francis Buchanan during his travels in Bihar in 1811–12. He reported a 'species of lameness', affecting villagers of all ages and both sexes, and resulting in muscular weakness in the legs and painful, irregular movement. But he dismissed the idea that *kesari* dal might be the cause of this disease as 'fanciful'.[86] Further notices of the disease soon followed. In 1839, the civil surgeon of Sarun district, Robert Rankine, remarked on the presence of the disease, attributing it to the extreme poverty of the labouring classes in the district and their 'complete dependence' on the zamindars.[87] A few years earlier, Colonel William Sleeman (better known for his role in the suppression of *thugi*), travelling in central India, also noted the paralysis caused by lathyrism. He linked it to the recent famine of 1833 and more specifically to consumption by the poor of a wild vetch, 'which though not sown of itself, is left carelessly to grow among the wheat and other grain'.[88] The most assiduous investigation of the disease was made in a series of remarkable reports in the 1850s and 1860s by James Irving, the civil surgeon at Allahabad. He gave a precise medical description of the 'palsy', prepared

---

[85] Wujastyk, *Roots*, 15.

[86] Francis Buchanan, *An Account of the Districts of Bihar and Patna in 1811–12* (2 vols, Patna: Bihar and Orissa Research Society, n.d.), 1: 274.

[87] Robert Rankine, *Notes on the Medical Topography of the District of Sarun* (Calcutta: G. H. Huttmann, 1839), 37.

[88] W. H. Sleeman, *Rambles and Recollections of an Indian Official* (2 vols, London: Hatchard & Son, 1844), 1: 134.

detailed tables showing the extent of its 'injurious effects' and its local distribution and was in no doubt that eating the 'poisonous vetch' was the cause. Through his enquiries, Irving was able to establish that villagers, while recognizing the connection between eating *kesari* dal and the onset of paralysis, saw no choice but to rely on it as the main item in their diet when other crops failed or they had no alternative means of subsistence, only hoping that they would not become permanently incapacitated by doing so. In other words, they took the risk of poisoning themselves rather than starve to death. Behind all this lay 'the poverty of the people'.[89]

Lathyrism became the subject of a remarkably detailed and enduring scientific investigation, stretching from the 1830s into the 1960s. Discussion of the disease and its connection with *kesari* dal continued throughout the colonial period and beyond, with controversy over its aetiology kept alive by claims that some other plant or ergot might perhaps be to blame.[90] In all of this extensive and technical literature, the underlying connection between poverty and poisoning was seldom lost sight of. The paralysis caused by lathyrism was particularly known to occur following periods of famine and food shortages, as during the late 1930s and again in the later stages of the Second World War and its aftermath, when food prices in northern and central India soared to exceptional levels.[91]

But *Lathyrus sativus* was not the only plant to fall under suspicion as being responsible for outbreaks of poisoning. Another was the millet *Paspalum scrobiculatum*. Known as *varagu* in Tamil and in Hindi-speaking regions as *kodo* (or *kodon*), the grain was cultivated as a staple crop in drier, un-irrigated areas. In Tamil-speaking south India, it was – and still is – prized as a tasty and nutritious alternative to rice. But from the 1860s and 1870s, as the scientific investigation of Indian diets gathered momentum, reports began to circulate in the Madras Presidency that *varagu* was a potentially dangerous crop that (at least during certain seasons or conditions of storage) caused illness and even death to humans and animals. It was even rumoured that *varagu* was so potent it could be

[89] James Irving, 'Report on a Species of Palsy Prevalent in Pergunnah Khyragurh, in Zillah Allahabad, from the Use of Kessaree Dall, as an Article of Food', in *Selections from the Records of Government, North-Western Provinces, Vol. 2* (Allahabad: Government Press, North-Western Provinces, 1866), 265–76; idem, 'Notice of Paraplegia Caused by the Use of *Lathyrus sativus* in the Various Districts of the North-Western Provinces of India', *IAMS* 12 (1868): 89–124.

[90] E.g., Andrew Buchanan, *Report on Lathyrism in the Central Provinces in 1896–1902* (Nagpur: Albert Press, 1908); Hugh Acton, 'An Investigation into the Causation of Lathyrism in Man', *IMG* 57 (1922): 241–47; L. A. P. Anderson, A. Howard and J. L. Simonsen, 'Studies on Lathyrism, I', *IJMR* 12 (1924–25): 613–44.

[91] Editorial, 'Lathyrism', *IMG* 74 (1939): 421–22; K. L. Shourie, 'An Outbreak of Lathyrism in Central India', *IJMR* 33 (1945–46): 239–48.

used to kill tigers.[92] At one stage, the Madras government toyed with the idea of banning *varagu* cultivation entirely, but this seemed utterly impractical and insufficient scientific evidence could be produced of its toxicity. In 1874, the chemical examiner reported on a sample of the grain that 'neither chemistry nor the microscope has been able to detect either poison or disease'.[93] The matter was then dropped, and perhaps we can regard this scare, unlike lathyrism, as more speculation than science. But reports continued to appear in the medical press of acute (but not fatal) poisoning from *kodon* flour.[94] The grain's hazardous nature was further highlighted towards the close of the Second World War when the Madras government, grappling with massive food shortages, encouraged the consumption of this and other millets. Laboratory tests showed it could be toxic to dogs.[95]

Poison – and an awareness or knowledge of the multifarious uses of poisonous substances – was widely disseminated in pre-colonial and early-nineteenth-century India. A notion of poisons and poisoning existed in India's high literary tradition – in its medical texts, especially – alongside a conviction that such poisons could be managed and either used therapeutically or their most damaging effects removed. Acquaintance with substances like opium, datura and *lal chitra* informed subaltern medical practice and was used in infanticide, suicide and abortion. But this is not to suggest that Indians had a complete and comprehensive mastery over poisons. They did not. The medical literature of the nineteenth century is replete with references to Indians who fell ill or died as a result not of homicidal poisoning with arsenic or aconitum but from the supposedly therapeutic use of dangerous drugs like nux vomica or oleander root in tonics, elixirs, aphrodisiacs and putative cures for venereal disease, or from accidental poisoning, caused by eating seeds, fruits and roots that they mistakenly thought to be harmless or which, from poverty and desperation, they took the calculated risk of eating.[96] India's

---

[92] William Robert Cornish, *Reports on the Nature of the Food of the Inhabitants of the Madras Presidency* (Madras: United Scottish Press, 1863), 7; G. D. Leman, Collector, Kistna, to Secretary, Madras, Board of Revenue, 21 August 1874, Madras, Board of Revenue, no. 2847, 2 October 1874, IOR.

[93] H. King, Madras, to District Surgeon, Salem, 22 September 1874, Madras, Board of Revenue, no. 3107, 27 October 1874, IOR.

[94] Anand Swarup, 'Acute "Kodon" Poisoning', *IMG* 57 (1922): 257–58.

[95] K. V. Sundara Ayyar and K. Narayanaswami, 'Varagu Poisoning', *Nature* 163 (1949): 912–13.

[96] T. Murray, 'Case of Poisoning from the Oleander Root', *IMG* 12 (1877): 319–20; Editorial, 'Strychnia Poisoning in India', *IMG* 20 (1885): 76–77; J. Venkata Swamy, 'Poisoning by Strychnos Nux Vomica', *IMG* 24 (1889): 113; P. Fitzpatrick, 'Case of Oleander Poisoning', *IMG* 24 (1889): 307; Chunilal Bose, 'On the Chemistry and Toxicology of *Nerium odorum*', *IMG* 36 (1901): 287–90.

poison culture was extensive and complex but also fragmented and contradictory. India lacked a more homogeneous and scientifically grounded understanding of what poisoning was or might be, and, more especially, how it could be governed and made subject to law. From early in the nineteenth century, that was precisely what colonialism began to construct.

2   The imperial *pharmakon*

One of the objectives of this book is to trace the emergence of toxicology as a body of scientific knowledge and practice in nineteenth- and twentieth-century India and to examine the 'conditions of possibility' that enabled or constrained its rise to a position of social, governmental and scientific authority.[1] While a wide and often-sophisticated knowledge of poisons was clearly present in pre-colonial India and was to some degree systematized through the indigenous medical systems, especially Ayurveda, it would be hard to speak of toxicology as a science dedicated to the study of poisons existing in India before the early decades of the nineteenth century. Until the 1830s, the study of poisons existed in a dependent relationship to something else – it was invoked to minister to specific medicinal needs, in the service of political assassination and dynastic ambitions or, at a more popular level, as an aid to homicide, infanticide and abortion. As both a process of knowledge transmission and institutional borrowing from the West and as a result of local influence, agency and necessity, toxicology in India began from the 1830s to acquire its own character, to accumulate its own arsenal of professional expertise and authoritative texts, approved techniques and effective methodologies.[2] Even while its ability to access evidence and to interpret it objectively and consistently remained problematic, toxicology gained recognition in colonial governance and public discourse. Toxicology continued to draw upon other sciences – chemistry and botany in particular – but it developed in different ways from those sciences in part because of its dual engagement with both public and state.

---

[1] Michel Foucault, *The Order of Things: An Archaeology of the Human Sciences* (London: Tavistock Publications, 1970), xxii.

[2] On the importance of the 1830s as a turning point in the investigation of poisoning and the development of toxicological expertise in England, see Katherine Watson, *Poisoned Lives: English Poisoners and Their Victims* (London: Hambledon Continuum, 2004), 169–70.

The making of toxic science

There are many routes into understanding the genealogy of what we might usefully think of as toxic science in India (encompassing both person-specific poisoning and the poisoning or pollution of the environment). One obvious point of entry is through substances. Many of the commodities that became categorized as toxic in British India occupied for much of the nineteenth century an ambiguous and shifting terrain – a pharmacological no-man's-land – between what was hailed as therapeutic and what was cautioned against as poisonous. This was an imperial version of the ancient notion of the *pharmakon*. In the original Greek, that term could mean either a medicine or a poison. It is now more commonly used to signify the simultaneous possibility of both curing and poisoning, a twilight zone where the healing art of the physician and the destructive guile of the poisoner might collide, overlap or become melded into one.[3]

The potential for curing and for causing injury coexisted at one and the same moment – depending on the strength of the dose administered, how it was applied and the physical state of the patient/victim. A drug derived its healing power, or its injurious properties, in part, too, from the intentionality of those – physician, poisoner or both simultaneously – who administered the substance, judged an appropriate dosage or the most effective moment for its delivery. But under empire the Jekyll-and-Hyde notion of the *pharmakon* extended beyond any one specific drug or the individual who administered it to the representation of medical systems at large. The *pharmakon*'s inherent ambiguity might be universal, but its configuration and effects were subject to the cultural and political domain within which substance and practitioner operated. In India's complex poison culture, and within the context of an evolving, empire-wide and increasingly global system of pharmacology and toxicology, a single substance could at one and the same time be identified as possessing dramatically different, even opposing, properties. What was deemed a poison when handled by an Indian hakim (presumed ignorant and careless) might be regarded as a potent and useful medicinal drug when entrusted to the safe hands of the European pharmacist. Some drugs, once the epitome of evil, lapsed into relative innocence, while others, once deemed relatively innocuous, or merely mildly narcotic, became so stigmatized as to be represented as an imminent danger and criminal threat to the patient/victim's physical and mental health.

---

[3] Jacques Derrida, *Dissemination* (London: Athlone Press, 1981), 70; Ian Burney, *Poison, Detection and the Victorian Imagination* (Manchester: Manchester University Press, 2006), ch. 2.

Several substances in particular became objects of intensive scientific scrutiny during the nineteenth century. The social uses of substances like opium, datura and aconite alluded to in Chapter 1 served as a local resource from which colonial botanists, biochemists and pharmacologists quarried medical knowledge but which they also used to differentiate between therapeutic drugs and dangerous poisons. The most important individuals engaged in this dual task of reconnaissance and ordering were medical men employed by the English East India Company and, following the uprising of 1857–58, by its successor, the British Crown, and served as members of the Indian Medical Service (IMS).[4] As state employees, their interests were determined as much by the needs of the regime they served as by their own scientific curiosity and professional ambitions. While regarding themselves as pioneers of a new scientific objectivity, they were necessarily influenced by the prejudices and presumptions of the colonial milieu in which they lived and worked. Practitioners of colonial medicine and science belonged to an epistemic community in which metropolitan precepts and practices mingled with local needs and indigenous ideas while also absorbing and addressing the requirements of an exogenous state. This is not to say that the colonial epistemic community was therefore homogeneous or unchanging, but it did share a common scientific and administrative culture, one that valued the local expertise acquired through long years of service in India, and, as can be seen most clearly in relation to medical jurisprudence (Chapter 4), for whom the idea that India was different in certain fundamental respects from the West was a strongly held conviction.

Venomous snakes

This chapter's primary concern is with poisons of plant origin. These were by far the most numerous substances identified as poisonous in India and, until late in the nineteenth century, they received the bulk of scientific attention. But plants (and minerals) were not the only sources of poisoning known to India or the only sites of toxicological enquiry. Snakes and snake venom occupy a conspicuous place in India's poison-lore and toxic histories. Serpents recur frequently in Hindu myth and

---

[4] The allusion here to 'medical *men*' is no accident. The formalization of toxicological knowledge by male physicians and pharmacists during the colonial period represented a long-term displacement and denigration of the poison knowledge and poison practices of women, as discussed in the previous chapter. On the IMS, see Mark Harrison, *Public Health in British India: Anglo-Indian Preventive Medicine, 1859–1914* (Cambridge: Cambridge University Press, 1994); Anil Kumar, *Medicine and the Raj: British Medical Policy in India, 1835–1911* (New Delhi: Sage, 1998).

Figure 1. The power of snakes 1: a *naga* (snake) stone near
Srirangapatnam, Karnataka, 2015.

legend and not only as expressions of evil or danger: the cobra that
spreads its protective hood to shade the infant Krishna from the heat of
the sun is indicative of a more benign role. In south India, carved *naga*
stones, especially in the form of entwined cobras, are widely venerated
and identified with human fertility (Figure 1). They are one of the ways in
which the beyond-human power of the venomous serpent is made visible
and incorporated into a culture of human anxiety and need. The

propitiation of snakes and belief in the snake goddess Manasa have long coexisted with a textual tradition that makes extensive reference to snake-bite and its treatment.[5] There was, besides, a conviction among many Ayurvedic practitioners (particularly the *vaids* or *kavirajs* of Bengal) that snake venom could be used to treat some of the most violent human disorders, such as cholera. This was a practice colonial physicians were inclined to scoff at, but it found its way into Indian homoeopathy as well as Ayurveda, and in the 1930s was the subject of renewed scientific investigation by the Indian pharmacologist R. N. Chopra.[6]

Beginning in the 1790s with Patrick Russell's remarkable descriptions of the snakes of the Coromandel Coast, the study of snakes and their venom formed a specialist field of zoological and toxicological enquiry in India.[7] As the illustrations in Russell's volumes also attest, scientifically reliable accounts of snakes called for skilled visual representation as well detailed textual description. They had the very practical aim of addressing the 'indiscriminate apprehension' that all Indian snakes were poisonous and the need to distinguish between venomous and non-venomous species.[8] The reasons for the wider concern are not hard to fathom, given the diversity of snake life in India and the great number of veno-mous cobras, kraits and vipers (Figure 2). Snakebite, a significant element in the human casualties admitted to hospitals and dispensaries, also caused the death of cattle and other livestock and so entailed sub-stantial loss to peasants and pastoralists. For Indians who laboured in the fields during the rainy season or ventured into jungles, poisoning was more likely to be encountered through accidental snakebite than almost any other cause. Snake poisoning seldom occurred 'in crowded cities with hospitals always open'. It mostly happened 'in the fields where the pea-sant steps on the lurking reptile, or more frequently still in the remote village where the wife is bitten in a dark corner of the hut by the snake which superstition has compelled her household to protect'.[9] As with

[5] Henry R. Zimmer, *Hindu Medicine* (Baltimore: Johns Hopkins University Press, 1948), 38–43; N. Subramanya Aiyar, 'Certain Facts Regarding the Poison-Lore of the Hindus', *IMG* 31 (1896): 5 8.

[6] J. Fayrer, 'Note on the Use of Snake-Poison in Medicine by the Koberajes of Bengal', *IAMS* 14 (1870): 226–31; Editorial, 'The Materia Medica of the Hindus', *IMG* 12 (1877): 190; Pratik Chakrabarti, *Bacteriology in British India: Laboratory Medicine and the Tropics* (Rochester, NY: University of Rochester Press, 2012), 136–40. For cobra venom in Indian homoeopathy, see Mahendralal Sircar, *A Sketch of the Treatment of Cholera* (2nd ed., Calcutta: P. Sircar, 1904).

[7] Patrick Russell, *An Account of Indian Serpents Collected on the Coast of Coromandel* (4 vols, London: G. Nicol, 1795–1809).

[8] Barbara J. Hawgood, 'The Life and Viper of Dr Patrick Russell: Physician and Naturalist', *Toxicon* 32 (1994): 1295–304.

[9] A. J. Wall, *Indian Snake Poisons, Their Nature and Effects* (London: W. H. Allen, 1883), vi.

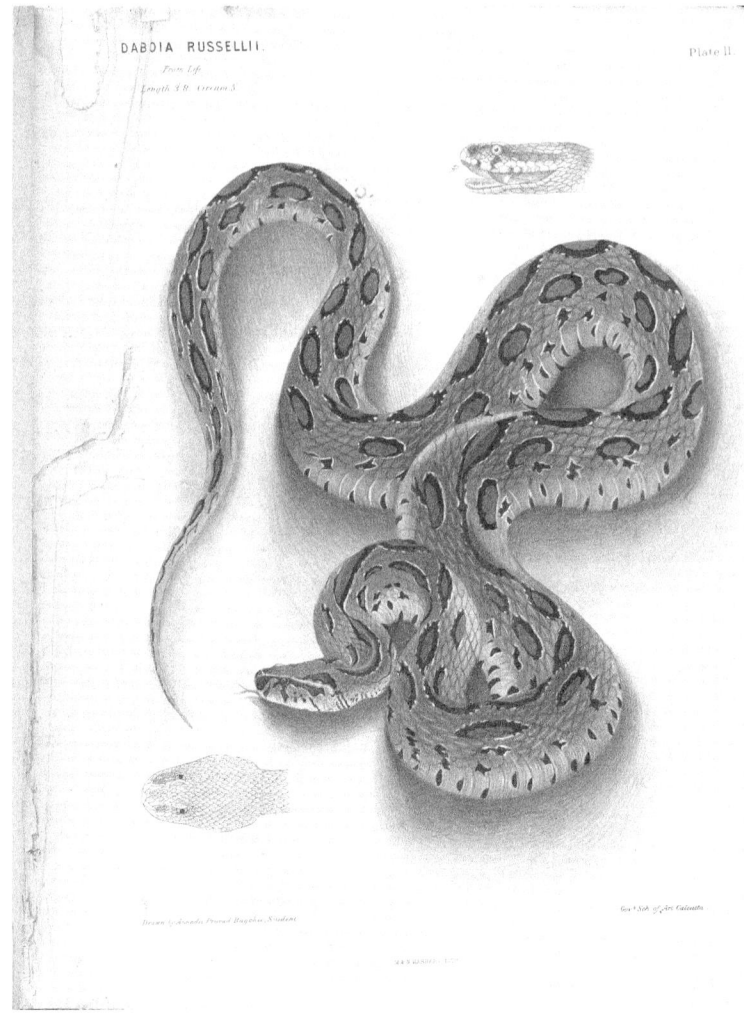

Figure 2. The power of snakes 2: illustration of a Russell's Viper from Joseph Fayrer, *The Thanatophidia of India* (1872), plate 12.

other forms of poisoning, the remoteness of much of India from medical aid of all but the most elementary kind made the danger of dying all the greater.

The official figures are staggering. It was estimated in the early 1870s that snakebite, mostly from cobras, caused more than 20,000 deaths a year in an Indian population of around 250 million.[10] Joseph Fayrer of the IMS campaigned for years and long after his retirement from India in 1872, for action to curb the mortality caused by poisonous snakes. By his reckoning, between 150,000 and 200,000 people perished from this 'evil' between 1870 and 1882.[11] He even proposed that a special government agency should be established, along the lines of the Thugi and Dacoity Department (see Chapter 3), to combat the persistent danger to human life from snakes, tigers and other wild animals.[12] This did not happen, but the Government of India did institute rewards for the destruction of poisonous snakes. In 1880, the Bengal government calculated that nearly 110,000 venomous snakes had been killed in that province alone since the introduction of rewards five years earlier. During that time, however, a further 51,000 people had died of snakebite. It was acknowledged that, even if the danger could be reduced, it would never be possible to eliminate poisonous snakes from India.[13] As late as 1901, 11,343 people and 850 cattle died from snakebite in the Bengal Presidency.[14]

While official efforts focused on the destruction of snakes in the wild, science in the laboratory concentrated on the investigation of snake venom and the search for effective antidotes. Here, again, Fayrer was a pioneer, conducting a series of experiments in Calcutta and London on the ways in which cobra venom attacked the human body.[15] Scientifically, too, the action of snake venom was important for the analogy it provided with other forms of poisoning, like that caused by the seeds of *Abrus precatorius* (see Chapter 6). Throughout the nineteenth century, and into the twentieth, the chemical composition of venom, its physiological effect and the antidotes and treatments suited to different snake species remained an important topic of medical research, providing material for dozens of articles, monographs and reports.[16] Snake farms were

---

[10] T. Lauder Brunton and J. Fayrer, 'On the Nature and Physiological Action of the Poison of *Naja tripudians* and Other Indian Venomous Snakes', *Proceedings of the Royal Society of London* 21 (1872–73): 359.

[11] J. Fayrer, 'Destruction of Life in India by Poisonous Snakes', *Nature*, 28 December 1882, 205–08.

[12] J. Fayrer, 'Destruction of Life in India by Wild Animals', *Nature*, 18 January 1883, 268–70.

[13] A. Smith, Commissioner, Orissa, to Secretary, Bengal, Judicial, 4 August 1879, and Horace A. Campbell, Secretary, Judicial, Political and Appointments, Bengal, to Secretary, Home, Revenue and Agriculture, India, 10 January 1880, Bengal, Judicial, no. 1, 16 January 1880, IOR.

[14] J. A. Bourdillon, Chief Secretary, Bengal, to Secretary, India, Home, 21 April 1902, Bengal, Judicial, nos 45–46, April 1902, IOR.

[15] Barbara J. Hawgood, 'Sir Joseph Fayrer: Snakebite and Mortality in British India', *Toxicon* 34 (1996): 171–82.

[16] *Report on the Effects of Artificial Respiration, Intravenous Injection of Ammonia, and Administration of Various Drugs &c. in Indian and Australian Snake-Poisoning* (Calcutta:

Figure 3. Snake venom as antidote and cure. 'Taking venom from a Russell's Viper', from *Seventh Congress of the Far Eastern Association of Tropical Medicine Souvenir: The Indian Empire* (1927).

established to harvest venom, and, like rabies, the investigation and treatment of snake poisons became a high priority for research in India's newly founded bacteriological institutes (Figure 3). The Haffkine Institute in Bombay, founded in 1899 as the Plague Research Laboratory, conducted extensive research on snake venom and kept

Bengal Secretariat Press, 1874); F. Wall, *The Poisonous Terrestrial Snakes of Our British Indian Dominions* (4th ed., Bombay: Bombay Natural History Society, 1928). On the scientific investigation of Indian snakes, see Chakrabarti, *Bacteriology*, ch. 4.

over 500 poisonous snakes to milk their venom and experiment with antidotes.[17] India's reputation as a place of poisons extended beyond human poisoners to the toxic lurking in wild nature. Occidental narratives and visual representations of India capitalized on this fear of venomous serpents or replayed the familiar trope of the Oriental 'snake charmer' who held command over them. And, as so often with poison histories, fantasy supplied what factuality might lack. Fertile Western minds imagined that snakes might become a deliberate instrument of murder, a technique so obvious, according to Sherlock Holmes, as to occur to any 'clever and ruthless man' who happened to have 'an Eastern training'.[18]

Snakes, as a source of poison but also of potential antidotes and remedies, had long been integral to India's indigenous medical systems and toxicological knowledge, and one can see science under the British as expanding and enlarging upon this pre-existing knowledge base even while developing new scientific techniques and research agenda and questioning claims made for snake venom's therapeutic properties. But the investigation of snake venom, although closely allied to other branches of medical science (bacteriology in particular), and frequently entrusted to the same IMS cadres as those who investigated other aspects of Indian toxicity, developed as a distinct branch of scientific enquiry, one largely free from the taint of criminal association that dominated so much colonial toxicology.

Founding forensic chemistry

The 1830s marked a watershed in colonial medical science as they did in many other aspects of British rule in India. In the Charter Act of 1833, the East India Company lost the last vestiges of the commercial role with which it had begun in 1600. After decades of annexation and conquest, dating back to the 1750s, the Company had consolidated its rule over a large part of the subcontinent and established much of the institutional and administrative structure by which British rule would be maintained until its demise in 1947. The relative cultural openness and social accommodation of earlier decades, though never more than partial, was

---

[17] *Haffkine Institute Platinum Jubilee Commemoration Volume, 1899–1974* (Bombay: Haffkine Institute, n.d.), 7–8.

[18] Arthur Conan Doyle, 'The Speckled Band', in *The Adventures of Sherlock Holmes* (London: Penguin, 1994), 200. This was fiction, but Fayrer cited an incident where two 'snake gurus' offered to teach a group of men 'incantations' that would protect them from snake poisoning. When the men refused to pay for their services, the 'gurus' unleashed several venomous snakes on them which chased them into a nearby field: two men died and a third was seriously injured in the attack: J. Fayrer, 'Deaths from Snake-Bites', *IMG* 5 (1870): 156–57.

replaced by a partisan, reformist agenda, symbolized by the abolition of sati by Lord Bentinck's administration in 1829. After years of experimenting with a pluralistic scheme of medical education in which instruction in Unani-tibb and Ayurveda coexisted uneasily with the teaching of allopathic medicine, in 1835 the Government of India established a medical college in Calcutta to patronize and propagate the Western system alone and train Indians in its use.[19] Equivalent medical institutions followed in the other provincial capitals – in Madras (now Chennai) in 1835 and in Bombay (at Grant Medical College) in 1845.

In Calcutta, the creation of a medical college was accompanied by the appointment of a professor of chemistry. The first holder of this post was W. B. O'Shaughnessy. Born in Ireland in 1809, he received his medical education at the University of Edinburgh and had already acquired a reputation as a forensic chemist before joining the Bengal Medical Service in 1833.[20] He is now remembered most for his experiments with Indian hemp (cannabis) in the late 1830s and, a decade later, for his role in overseeing the creation of the Indian telegraph system.[21] But at the time of his appointment to Calcutta, O'Shaughnessy was a lowly assistant surgeon, languishing with an up-country regiment in Muttra. He was, however, determined to impress on government the paramount importance of chemistry (and therefore toxicology and medical jurisprudence) to the teaching and practice of medicine. As he explained to the Governor-General in March 1835, this was partly from his warm personal attachment to chemistry, a subject that had long been his 'constant and favourite study'. But he also regarded its teaching and dissemination as 'a paramount necessity' for India. In no other country was there a richer field for research, or one where the needs of medicine and the progress of the inhabitants so demanded its pursuit, and yet in which it remained so 'unaccountably neglected'.[22] O'Shaughnessy's ambition did not go unrealized. Part of the importance of toxicology in India was the opportunity it provided for chemistry (as, latterly, for biochemistry and bacteriology) as a field of scientific endeavour and for the application of science to both state and public needs.

[19] David Arnold, *Colonizing the Body: State Medicine and Epidemic Disease in Nineteenth-Century India* (Berkeley: University of California Press, 1993), 54–58; Kumar, *Medicine*, 19–36.

[20] M. J. Bramley, Superintendent, Medical College, Calcutta, 9 July 1835, India, Public Consultations, no. 14A, 5 August 1835, IOR.

[21] James H. Mills, *Cannabis Britannica: Empire, Trade, and Prohibition, 1800–1928* (Oxford: Oxford University Press, 2003), 39–46; Deep Kanta Lahiri Choudhury, 'Beyond the Reach of Monkeys and Men? O'Shaughnessy and the Telegraph in India, c. 1836–56', *Indian Economic and Social History Review* 37 (2000): 331–59.

[22] W. B. O'Shaughnessy to Lord Metcalfe, 31 March 1835, India Public Consultations, no. 15, 5 August 1835, IOR.

Before arriving in India, O'Shaughnessy had written a commentary on Robert Christison's *Treatise on Poisons* and carried a letter of recommendation from this celebrated toxicologist with him to India.[23] In Calcutta he set up his own chemistry laboratory (the first of its kind in India) and conducted further experiments, proposing improvements, for example, to the famous 'Marsh test', instituted by James Marsh at the Royal Arsenal in Woolwich for the detection of arsenic.[24] At the request of the Calcutta coroner and police department, he prepared reports on the stomach contents and viscera of individuals suspected of having died from poisoning by arsenic and orpiment (arsenic trisulphide). He investigated *bish* and *lal chitra* and described their use as poisons and abortifacients. He differentiated between the effects of arsenic poisoning and the symptoms of cholera, between which there was (and remained) widespread confusion.[25] In his *Manual of Chemistry*, compiled for students at the medical college, O'Shaughnessy wrote at length on the subject of poisons, arguing for the importance of chemistry in detecting criminal poisoning and so as a means to 'bring the murderer to justice ... and protect innocent persons labouring under false accusations'.[26]

The fruits of O'Shaughnessy's enquiries, backed by laboratory experiments (see Figure 4), post-mortem analyses, conversations with *vaids* and familiarity with local medical practice as well as metropolitan medical jurisprudence, were incorporated into his *Bengal Dispensatory* and *Bengal Pharmacopeia*.[27] But, despite the international recognition he gained for his work on medical jurisprudence and toxicology, in the eyes of the government, O'Shaughnessy's most significant contribution lay elsewhere – in the application of chemistry to the needs of trade and industry (such as assessing the suitability of Indian clays for making firebricks and other ceramic goods currently imported at great cost). Even in his college lectures, O'Shaughnessy spoke on such mundane topics as dyeing, bleaching, distilling, the smelting of ores and the manufacture of drugs. It was primarily for this work, rather his medico-legal contributions, that O'Shaughnessy was given the additional title (and extra remuneration) of

---

[23] For Christison's importance, see Burney, *Poison*, 41–43.

[24] W. B. O'Shaughnessy, 'On the Detection of Arsenic Poisons by Marsh's Process', *JASB* 8 (1839): 147–49.

[25] W. B. O'Shaughnessy, *Report on the Investigation of Cases of Real or Supposed Poisoning* (Calcutta: Bishop's College Press, 1841).

[26] W. B. O'Shaughnessy, *Manual of Chemistry* (1842), cited in Mel Gorman, 'Sir William Brooke O'Shaughnessy, Anglo-Indian Forensic Chemist', *Notes and Records of the Royal Society of London* 39 (1984): 59.

[27] W. B. O'Shaughnessy, *The Bengal Dispensatory* (Calcutta: William Thacker, 1842), 165–67; idem, *The Bengal Pharmacopoeia* (Calcutta: Bishop's College Press, 1844).

Figure 4. W. B. O'Shaughnessy, Professor of Chemistry and Natural Philosophy, Medical College Calcutta.

Chemical Examiner in June 1840.[28] Other provinces gradually followed Bengal's lead in appointing their own chemical examiners (known in Bombay as the Chemical Analyser).

[28] Bramley, 9 July 1835, India Public Consultations, no. 14A, 5 August 1835, IOR; India, Public letter, 3 June 1840 to Court of Directors, London, F/4/1957: 85390, IOR.

O'Shaughnessy was the pioneer of forensic chemistry in India.[29] His science is well illustrated by the account he published in the 1841 of nux vomica, which shows the author's careful and methodical approach and his awareness of the hazards and opportunities presented by India's pharmakonic substances. He began by recording indigenous names for the poison-bearing nut, and he noted that the tree that bore it, *Strychnos nux-vomica*, was native to the Coromandel Coast and the jungles of Bengal, citing his own research to prove this was the source of the drug sold in Indian bazaars. He reported on the analysis of the nut's chemical composition by the French chemists Joseph Pelletier and J.-B. Caventou, who in 1818 had identified the toxic principle as strychnine and, in the following year, revealed a second alkaloid, brucine.[30] O'Shaughnessy cited various European authors to show how deadly strychnine could be: it was 'so poisonous that half a grain destroys a rabbit in five minutes in violent paroxysms of tetanus' (a reference to the disease whose symptoms it most closely resembled).[31] To these metropolitan sources, he added his own acquaintance with the drug in India:

When taken in very small or medicinal doses, all this class [of drug, i.e., strychnine] agree in proving at first stimulant and tonic, the appetite is improved, the powers of digestion increased. The secretions are rendered more copious and regular, sexual feelings are unusually excited, and cheerfulness of spirits often induced.[32]

If the dosage were increased, or the treatment persisted in, the patient began to experience more alarming symptoms – tremors, cramp, a prickling or creeping sensation and ultimately 'tetanic spasms'. He noted, however, a report by a Calcutta surgeon, that some Indians

take the kuchila nut [nux vomica] morning and evening, continuously for many months, beginning with an eighth of a grain, and gradually increasing the dose to an entire nut, or about 20 grains. If taken immediately before or after meals no unpleasant effect is produced, but if this precaution be neglected spasms are apt to ensue.[33]

From a diagnostic and forensic perspective, the symptoms of strychnine poisoning were worryingly indeterminate. The morbid appearances

---

[29] A further example of O'Shaughnessy's skill can be seen in his analysis of Indian 'hill opium': W. B. O'Shaughnessy to Secretary, India, Home, 21 August 1848, F/4/2401: 129687, IOR.

[30] Sacha Tomic, 'Alkaloids and Crime in Early Nineteenth-Century France', in José Ramón Bertomeu-Sánchez and Agustí Nieto-Galan (eds), *Chemistry, Medicine, and Crime: Mateu J. B. Orfila (1787–1853) and His Times* (Sagamore Beach: Science History Publications, 2006), 261–92.

[31] W. B. O'Shaughnessy, *Bengal Dispensatory and Pharmacopoeia* (Calcutta: Bishop's Press, 1841), 438.

[32] Ibid., 439.   [33] Ibid., 439–40. An apothecary's grain is equivalent to 65 milligrams.

of the body in fatal cases were 'variable and inconclusive'. In some instances, post-mortem analysis showed that

portions of the cerebellum and spinal cord have undergone softening. In others, these organs are unusually vascular, in many altogether unchanged. A few instances are on record in which the stomach and small intestines exhibited signs of violent irritation and inflammation. In some individuals the body becomes rigid immediately after death, and seems as if it were fixed in the tetanic spasms.[34]

Yet, for all strychnine's toxic dangers, O'Shaughnessy saw clear benefits in its judicious medicinal use. It could be administered to treat local and general paralysis, rheumatism, neuralgia and intermittent fever, as a means of preventing incontinence and (perhaps with the Indian use primarily in mind) as 'a stimulant in sexual impotency'. In general, 'the physician possesses few remedies of greater value, none of superior power, and scarcely any so formidable' as strychnine afforded. But the 'utmost skill' was required of the doctor in deciding how such a fearsome drug should be administered, and whether it should be used at all. If the patient's symptoms became too violent, then rapid recourse had to be made to remedial measures – a large dose of opium or cannabis extract, leeching to the temples, a blister to the spine, a warm bath and enemata, or, failing those, a stomach pump.[35] Strychnine was not for the faint-hearted.

Bioprospecting

O'Shaughnessy demonstrated the value of chemistry to toxicology and medical jurisprudence in India, but in this he had few immediate successors. In 1843, he was succeeded as Chemical Examiner by Frederic J. Mouat, who, as student in Paris and before joining the medical service in India, had attended lectures on medical jurisprudence given by Mathieu Orfila, 'that eminent continental Toxicologist'.[36] Shortly after becoming Chemical Examiner, Mouat called for an enquiry into dangerous drugs and their criminal uses. He estimated that there were at least three dozen such drugs sold openly in the bazaars of Bengal, emphasizing the hazards posed by the bazaar as India's main – and unregulated – medical marketplace. As well as vegetable poisons, Mouat noted with alarm the sale of mineral poisons, including some that were as yet undetectable by chemical science. Concerned at the 'large amount of crime which can thus be committed with impunity throughout the country', he

---

[34] Ibid.   [35] Ibid., 441.
[36] Frederic J. Mouat, *Report on Jails Visited and Inspected in Bengal, Behar, and Arracan* (Calcutta: Military Orphan Press, 1856), 129.

regarded such unlicensed sales as a matter of 'very great importance to the state'.[37] Here was a clear statement of the perceived political importance of toxicology, even though little data existed at the time to suggest that criminal poisoning was widespread. Mouat was not alone in his concern. In 1848, R. H. Irvine, Patna's Civil Surgeon, also complained about the unregulated sale of dangerous drugs by *pansaris* in the local bazaars, claiming that 'the quantity of arsenic and other poisons sold is terrible to think of'.[38] But, despite such expressions of alarm, nothing was done at the time to check the sale of dangerous drugs. Like O'Shaughnessy before him, Mouat was then moved onto another, apparently more urgent, posting. Where O'Shaughnessy was diverted onto telegraphs, Mouat became Inspector-General of Prisons for Bengal.

While the chemical, laboratory-based approach to toxicology stalled, other avenues of exploration thrived. The most significant of these was botany, a field in which the medical officers, or 'surgeons', of the East India Company excelled. From early in the nineteenth century, as part of their wider investigation of the natural history of India, the Company's surgeon-botanists developed their own, often censorious, knowledge of India's plant poisons and their local uses, and in so doing opened a new chapter in the social life of its toxic substances.[39] There is a significant point of contrast here. In many early-nineteenth-century works of medical jurisprudence produced in Europe and North America, attention was concentrated on mineral poisons rather than vegetable ones.[40] Robert Christison lavished more than 500 pages of his *Treatise on Poisons* on a detailed discussion of the toxic properties of arsenic, mercury, copper, antimony and lead before turning to the subject of opium. Other organic poisons of keen interest to India – datura, *Aconitum ferox*, nux vomica – were also relegated to the closing chapters.[41] In India, the reverse was the case, with vegetable poisons attracting, for most of the nineteenth century, greater concern than mineral ones.

Two reasons can be adduced for this. One was that in the nineteenth century, India was held to be one of the 'greatest drug-yielding regions in

[37] F. J. Mouat to Secretary, Home, 18 May 1843, Bengal, Public, no. 26, 12 June 1843, IOR.

[38] R. H. Irvine, *A Short Account of the Materia Medica of Patna* (Calcutta: Military Orphan Press, 1848), 2.

[39] This enquiry extended into the Malay Peninsula, where Malays, 'like other Eastern peoples', were said to be 'skilled in the art of poisoning': John D. Gimlette, *Malay Poisons and Charm Cures* (3rd ed., Kuala Lumpur: Oxford University Press, 1971), 1.

[40] In England, between 1750 and 1914, organic substances like opium and strychnine accounted for only a third of criminal poison cases, and so the story of poisoning was, in effect, 'a chronicle of the rise and fall of arsenic': Watson, *Poisoned Lives*, 32–33.

[41] Robert Christison, *A Treatise on Poisons* (4th ed., Philadelphia: Edward Barrington & George D. Haswell, 1845).

the world', rich in herbal medicines and poison-bearing plants.[42] The pharmacological texts of the period were overwhelmingly compendia of vegetable drugs, with relatively few notices of mineral or animal substances. The 'abundance of deadly plants' was cited as one reason why criminal poisoning in India was both widespread and difficult to suppress.[43] The contrast between nineteenth-century materia medica manuals and earlier Ayurvedic and Unani medical texts is also striking. Like the claims made for the therapeutic properties of snake venom, once-prized items of materia medica, like crushed pearls and bezoar stones or the flesh of owls and tigers, were now treated by sceptical surgeons as devoid of curative powers and unworthy of scientific consideration.[44] In diet, in medicine, in its toxic culture, India was seen as an overwhelmingly vegetarian society, and in this lay a further paradox. In the minds of the meat-eating British, Indian vegetarian diets, especially the rice diets of the east and south, were nutritionally deficient and a cause of the apparent frailty of Bengalis and Madrasis, in contrast even with the meat and wheat diets of the 'martial' Rajputs, Sikhs and Muslims of the north and north-west.[45] But if, with respect to diet, India's vegetarian civilization connoted weakness, with regard to its vegetable poisons it signified exceptional potency.

A second reason why plants figured so conspicuously in colonial pharmacology and toxicology was that Company surgeons were trained in botany as part of their medical education in Britain, especially at the Scottish universities, before travelling to India. O'Shaughnessy was unusual in combining medicine with chemistry, though even his *Bengal Pharmacopoeia*, published a decade after his arrival in Calcutta, inclined heavily towards medicinal plants. The apprentice surgeon's early induction into a utilitarian as well as aesthetic appreciation of plant life made botany an attractive professional and recreational pursuit in India, with

---

[42] F. H. Brett, *A Practical Essay on Some of the Principal Surgical Diseases of India* (1840), cited in Girindranath Mukhopadhyaya, *History of Indian Medicine from the Earliest Ages to the Present Time* (2 vols, Calcutta: University of Calcutta, 1923), 1: 92.

[43] Theodric Romeyn Beck and John R. Beck, *Elements of Medical Jurisprudence* (5th ed., London: Longman, Rees, Orme, Brown, Green & Longman, 1836), 662; A. H. Giles, 'Poisoners and Their Craft', *CR* 81 (1885): 93.

[44] George Playfair, *The Taleef Shereef, or Indian Materia Medica* (Calcutta: Medical and Physical Society of Calcutta, 1833); Francis Zimmermann, *The Jungle and the Aroma of Meats: An Ecological Theme in Hindu Medicine* (Berkeley: University of California Press, 1987).

[45] On Indian diets, see William Robert Cornish, *Reports on the Nature of the Food of the Inhabitants of the Madras Presidency* (Madras: United Scottish Press, 1863); A. H. Church, *Food-Grains of India* (London: Chapman & Hall, 1886), 75–76. Such criticisms did not emanate from Europeans alone: see Chunilal Bose, *Food* (Calcutta: University of Calcutta, 1930).

many surgeons attaining greater fame as botanists than they ever did as physicians.[46]

Bioprospecting, to use Londa Schiebinger's insightful term, can be understood as the empirical and epistemological process by which Western powers and their imperial agencies appropriated local knowledge of extra-European plants and incorporated them into their own commercial, medical and scientific systems.[47] But it is also possible to envisage another kind of bioprospecting in which Western toxicology investigated the properties and uses of the poisonous plants of Asia, Africa and the Americas. In this, there might be little commercial benefit, though British India did command a minor trade in several toxic substances and certainly a major one in opium. But toxicological enquiry might aid science by advancing botanical knowledge and support colonial governance by yielding information of value to criminology and medical jurisprudence. Local notices, like O'Shaughnessy's on nux vomica, were incorporated into works of medical jurisprudence in Europe and America, and physicians with Indian experience, like J. F. Royle, whom we will consider shortly, returned to Britain bringing their familiarity with Indian drugs and therapeutic techniques with them.[48] Nor could the possibility be ignored that even the most formidable poison might yield a useful medicine. Indeed, it was widely believed that, given the rapid and violent manner in which tropical diseases developed, heroic doses of strong medicines were needed to arrest their terrifying progress.[49] Only the exceptional power of a poisonous drug might be sufficient to effect a cure.

Taming *Aconitum*

One point of departure for the scientific investigation of Indian poisons was a catalogue of medicinal plants and drugs compiled by John Fleming, a Company surgeon-botanist, in 1810. Fleming sought to acquaint physicians newly arrived in India with the Sanskrit and Hindustani names for local and imported drugs and with their properties. Most of the substances Fleming noted were already known to the West: he made no reference at all to Indian aconites, and his brief entry on datura primarily referred to European experience, though he did

---

[46] Ray Desmond, *The European Discovery of the Indian Flora* (Oxford: Oxford University Press, 1992); David Arnold, *The Tropics and the Traveling Gaze: India, Landscape, and Science, 1800–1856* (Seattle: Washington University Press, 2006).

[47] Londa Schiebinger, *Plants and Empire: Colonial Bioprospecting in the Atlantic World* (Cambridge, MA: Harvard University Press, 2004).

[48] For example, Beck and Beck, *Elements*, 930–33.

[49] [Anon.], *Autobiography of an Indian Army Surgeon* (London: Richard Bentley, 1854), 76–77.

add, without much evident perturbation, that the 'soporiferous and intoxicating qualities of the seeds are well known to the inhabitants [of India]'. At worst, the seeds had the reputation for being used for the same 'licentious and wicked purposes' as noted by earlier European botanists in the East.[50]

There is little suggestion, then, in 1810 that India's plant drugs constituted a distinctive or particularly worrying source of toxicity, and yet this was precisely the time at which magistrates and police officers in northern and central India were becoming aware of the use of datura to poison travellers and hence of a connection between this common drug and the murderous 'cult' of *thugi*.[51] However, within a few years of Fleming's inventory, colonial botanists began to take India's poisonous plants more seriously. Among the earliest objects of toxicological enquiry were the several varieties of aconites imported into northern India from Nepal and the Himalayas, regions in the early nineteenth century that were still largely inaccessible to Europeans. In an account of his visit to Nepal in 1802–03, Francis Buchanan described the various species of aconites that grew in that mountainous country and which were believed to be the source of the substance known as *bish* or *bikh* (i.e. as '*the* poison'). Buchanan had great difficulty in distinguishing between the various aconites, with their almost identical roots, and the different medicinal or toxic properties attributed to each. For him, this was indicative of a greater confusion in India generally over plant taxonomy. Among the Hindus, he complained, botanical nomenclature was 'miserably defective, and can scarcely fail to be productive of the most dangerous mistakes in the practice of medicine'. Unable to visit Tibet in person, he sent a collector there to gather poisonous aconites only for his agent to return instead with 'Nirbishi, or [the] kind used in medicine'.[52]

Taxonomic uncertainty might abound, but one of these species, which he identified as *Aconitum ferox*, Buchanan called 'this dreadful root' and claimed that 'large quantities' of it were imported annually into India. The root was, he reported, 'equally fatal when taken into the stomach, and applied to wounds, and is in universal use throughout India for poisoning arrows'. It was also reputedly used for 'the worst of purposes' (i.e. murder) as well. In probably the first proposal of its kind, he

---

[50] John Fleming, 'A Catalogue of Indian Medicinal Plants and Drugs', *AR* 11 (1810): 165–66.

[51] Kim A. Wagner, *Thuggee: Banditry and the British in Early Nineteenth-Century India* (Basingstoke: Palgrave Macmillan, 2007), 47, 60, 71. For datura *thugi*, see Chapter 3.

[52] Francis Hamilton [Buchanan], *An Account of the Kingdom of Nepal* (Edinburgh: Archibald Constable, 1819). Buchanan took the surname Hamilton after leaving India in 1815.

suggested that the importation of this highly toxic and dangerous substance should be subject to 'the attention of the magistrate' – that is to say, brought under government surveillance, if not total prohibition.[53] Uncertain though Buchanan's identification of the offending *bish* was, his dire warning of its criminal dangers and the labelling of the plant as *ferox* (ferocious) fostered the idea that this was one of the most lethal plant substances known to India, far worse than aconites, such as wolf's bane and henbane, known in Europe.

Later writers enlarged on the evil reputation of *A. ferox* and its frightening toxicity. In 1830, Nathaniel Wallich, the Danish naturalist who followed Buchanan as superintendent of Calcutta's botanic garden, endorsed his predecessor's views and declared that the plant yielded 'probably the most deleterious vegetable poison of continental India'.[54] Wallich included a colour plate of what he claimed was *Aconitum ferox* in his *Plantae Asiaticae Rariores*, though later authorities suspected that the root shown actually belonged to another, less toxic, species (see Figure 5).[55] Making a substance visible – in this case literally, by means of botanical drawings – did not necessarily ensure the reliability of that knowledge.

During his career in the Bengal Medical Service from 1819 to 1831, John Forbes Royle served as superintendent of the Company's other main botanic garden, at Saharanpur in north India, and wrote widely on Indian economic and medical botany.[56] In 1839, he published a definitive account of Himalayan flora, in which, despite the unresolved taxonomy of the aconites, he confirmed Buchanan's verdict on the lethal toxicity of *A. ferox* and its identity as *bish*, the quintessential Indian poison. Royle went on to denounce Indian physicians for their use of this and other 'powerful drugs, such as arsenic, nux vomica, and croton'. What began as the scientific account of a single plant escalated into an attack on 'native' physicians in general. Since the order of Ranunculaceae to which the aconites belonged stood first in Royle's flora (as it did by botanical convention in others), this outburst set the tone for his subsequent remarks on the unreliable nature of indigenous plant knowledge and the perils of *bish* in the hands of India's ignorant druggists and unqualified doctors.[57] Royle's medico-botanical work is germane to this discussion

---

[53] Ibid., 98–99.

[54] Nathaniel Wallich, *Plantae Asiaticae Rariores* (3 vols, London: Teuttel, Würtz & Richter, 1830–32), 1: 35. Wallich's description was widely reproduced: e.g. Beck and Beck, *Elements*, 920–21.

[55] Edmund John Waring, *Pharmacopoeia of India* (London: W. H. Allen, 1868), 3.

[56] On Royle's career, see Arnold, *Tropics*, 162–66.

[57] J. Forbes Royle, *Illustrations of the Botany and Other Branches of the Natural History of the Himalayan Mountains, and of the Flora of Cashmere* (2 vols, London: W. H. Allen, 1839), 1: 48.

Figure 5. Aconitum ferox, original watercolour by Vishnuprasad, reproduced as a lithograph in Nathaniel Wallich, *Plantae Asiaticae Rariores* (London, 1830), vol. 1, plate 41.

for another reason, too. The early nineteenth century witnessed an explosion of scientific knowledge about the chemical composition of medicinal drugs and poisons. One of the most momentous of these discoveries came in 1820 when Pelletier and Caventou succeeded in isolating quinine from cinchona bark, thereby paving the way for the widespread use of quinine in malaria treatment and prophylaxis. Surgeon-botanists like Royle were aware of these advances in biochemistry, citing them in their work, but, with little biochemical research conducted in India at the time (apart from O'Shaughnessy's), their reporting was mostly second-hand. In the case of *Aconitum ferox*, the discovery in Europe in 1833 of the highly toxic alkaloid *aconitine* or *aconitia* appeared to endorse the Indian drug's already formidable reputation.

On his return to England, Royle was appointed Professor of Materia Medica and Therapeutics at King's College London, a post that enabled him to combine his India expertise with the latest European developments in pharmaceutical chemistry. In the textbook he compiled during his London years, Royle again turned his attention to the aconites and highlighted the importance of *Aconitum ferox*, samples of which he had supplied for chemical analysis in Britain. Tests showed the Indian plant to contain three times as much *aconitia* ('the most poisonous of all known substances') as the English species, *A. napellus*. Although *aconitia* in general was classed as an extremely powerful poison, the exceptional toxicity of the Himalayan plant, *A. ferox*, made it a superior source of the drug, and, following Royle, the alkaloid extracted from the Indian root was particularly recommended for use in treating neuralgia, rheumatism and heart disease.[58]

Even so, ambiguity remained. In his *Cyclopaedia of India*, Edward Balfour named several Himalayan aconites but singled out *A. ferox* as the source of the 'formidable' poison *bish*, one-tenth of a grain of which was enough to kill a goat.[59] Citing earlier medical and botanical authorities, Balfour described it as 'one of the most celebrated articles in Indian medicine and toxicology', its roots being 'equally fatal taken internally or applied to wounds'. He repeated reports that it had once been used to kill wild animals and repel invading armies: 'The Gurkhas say that they could so infect all the water with the dreadful root that no enemy could advance into their mountain fastnesses.'[60] He further noted that the roots of different aconites, from the highly toxic to the simply 'inert', were barely distinguishable from each other, especially when sold in a desiccated state in the bazaars without other evidence as to their identity. Thus *Aconitum heterophyllum*, also native to the Himalayas, had a root almost identical to that of *A. ferox*, 'composed of two oblong tubers, of a light ash colour externally, white internally, and of . . . bitter taste'. It, too, acted 'as a bitter

---

[58] J. Forbes Royle, *A Manual of Materia Medica and Therapeutics* (3rd ed., London: John Churchill, 1856), 270–71; G. Lathom Browne and C. G. Stewart, *Reports of Trials for Murder by Poisoning by Prussic Acid, Strychnia, Antimony, Arsenic, and Aconita* (London: Stevens & Sons, 1883), 570, 575, 578.

[59] Edward Balfour, *The Cyclopaedia of India* (3rd ed., 3 vols, London: Bernard Quaritch, 1885), 1: 19.

[60] Ibid. The reference is to the British invasion of Nepal during the Anglo-Gurkha War of 1814–15.

tonic and febrifuge' and was used 'by Europeans and natives in Indian medicine as a tonic and aphrodisiac'.[61]

The surgeon-botanists' hostile view of *Aconitum ferox* gained added weight from its criminal associations. In 1890, William Dymock closed the entry on aconites in his *Pharmacographia Indica* by observing that 'Cases of accidental poisoning by aconite are occasionally met with arising from the use of the drug by ignorant native doctors or as a remedy for fever, &c.' He noted that 'Homicidal and suicidal cases are occasionally reported', but conceded that these were 'not so frequent as one might expect, considering how readily the drug can be obtained, and how well known are its poisonous properties'.[62] In the nineteenth century, aconitum was routinely listed alongside datura, opium and arsenic as one of a quartet of deadly Indian poisons, though its actual use in murder or suicide appears to have been rare. Of 366 suspected poisonings investigated by UP's Chemical Examiner in 1911, only four involved aconite.[63] Moreover, while repeating the convention that *A. ferox* was 'highly poisonous', it remained in use by European doctors in India as a tonic, antiperiodic and aphrodisiac.[64]

In 1889, as part of his enquiry into the economic products of India, George Watt called for 'a thorough investigation' of the aconites, 'with a view of establishing a trustworthy supply of uniform quality for medicinal purposes, and, if possible, of checking the indiscriminate way in which the drug is placed within the reach of persons desiring to use it for criminal purposes'.[65] He observed that the 'poisonous forms' had never been adequately identified, with the result that 'of a given weight of the root sold in our druggists' shops, a certain percentage frequently contains no aconitia whatever; indeed, an entire consignment may be perfectly inert'. Since it remained 'the most valuable drug known' for treating skin diseases, Watt added, it was 'very much to be regretted that so valuable a medicine should suffer in consequence of ignorance'.[66] It was only in 1905 that Otto Stapf at Kew Gardens in London produced a definitive

---

[61] Ibid.

[62] William Dymock, *Pharmacographia Indica: A History of the Principal Drugs of Vegetable Origin, Met With in British India* (3 vols, Calcutta: Thacker, Spink, 1890–91), 1: 8.

[63] *ARCE (UP), 1911*, 5. T. E. B. Brown, *Punjab Poisons* (3rd ed., Lahore: 'Civil and Military Gazette' Press, 1888), ranked aconite among India's commonest poisons, but cited only eight cases for the years 1879–86 compared to 1,286 for arsenic and 350 for opium over the period 1861–66 and 1879–87: ibid., iv, 130–37. Aconite poisoning was rare in Britain: see Browne and Stewart, *Reports of Trials for Murder by Poisoning by Prussic Acid, Strychnia, Antimony, Arsenic, and Aconita*, ch. 10.

[64] Waring, *Pharmacopoeia*, 3; William Owen, 'Report on the Treatment of Acute Dysentery by Aconite', *IMG* 17 (1882): 90–95.

[65] George Watt, *A Dictionary of the Economic Products of India* (6 vols, Calcutta: Superintendent, Government Printing, India, 1889–92), 1: 85.

[66] Ibid.

account of the Indian aconites, eliminating several of the twenty-two supposed species and distinguishing between the poisonous and non-poisonous species.[67] But, even if *Aconitum ferox* no longer commands such alarm as it did in nineteenth-century India, its controversial career is still not over: doubts continue to be raised about its safe use in present-day Ayurvedic and homoeopathic remedies.[68]

Demonizing datura

Datura was another Indian plant that came to be regarded as dangerous. It belongs to the same botanical family, the Solanaceae, that includes deadly nightshade and other atropine-yielding poisonous plants but also such wholesome foods as aubergines, potatoes and capsicums. In recent centuries, datura has been widely distributed across India, but there is uncertainty among botanists as to whether the various species and sub-species of datura are native to the region. Linnaeus' name for the plant, from the Sanskrit *dhatura*, hints at an antique presence in South Asia, and early textual references and widespread medicinal use across the region, including among tribal populations like the Santhals, suggest that some datura species have grown there for a long time.[69] When it came to poisoning, *stramonium* was the species most usually cited for its toxicity.[70] Common in gardens, along roadsides and on waste ground across large parts of India, datura was seldom thought worthy of much attention in nineteenth-century botanical texts. A rare illustration of datura from south India in 1845 bears the caption 'very common … springs up on rubbish, and seems to be one of those plants which follow man' (see Figure 6).[71] This rather disparaging view contrasts with datura's pre-colonial reputation, for (as indicated in Chapter 1) it was valued as being of wide medicinal use, including in the treatment of headaches, asthma and epilepsy.[72] But datura was also associated with individuals

[67] Otto Stapf, *Annals of the Royal Botanic Garden, Calcutta, Vol. X, Part II: The Aconites of India* (Calcutta: Bengal Secretariat Press, 1905).

[68] Ashok Kumar Panda and Saroj Kumar Debnath, 'Overdose Effect of Aconite-Containing Ayurvedic Medicine', *International Journal of Ayurveda Research* 1 (2010): 183–86.

[69] S. K. Jain and S. K. Borthakur, 'Solanaceae in Indian Tradition, Folklore, and Medicine', in William G. D'Arcy (ed.), *Solanaceae: Biology and Systematics* (New York: Columbia University Press, 1986), 577–83.

[70] R. N. Chopra, R. L. Badhwar and S. Ghosh, *Poisonous Plants of India* (2 vols, New Delhi: Indian Council of Agricultural Research, 1965), 2: 644–50.

[71] In Royle's *Illustrations* (1: 279), datura was included in discussion of the Solanaceae, but in his later *Manual*, 573–75, it appeared as 'an energetic poison'.

[72] Fanny Parks, *Wanderings of a Pilgrim in Search of the Picturesque* (2 vols, London: Pelham Richardson, 1850), 1: 148.

who were 'dissipated and depraved' and with those 'debauched devotees' of Shiva who craved 'delirious stupefaction'.[73] But, at worst, it was considered capable of causing delirium rather than death.

European awareness of datura in India extended as far back as the Portuguese physician Garcia da Orta in sixteenth-century Goa, who wrote that 'Those who take this medicine lose their heads'. They laughed and became 'very liberal', and in such a state of intoxication were easily hocussed or robbed. Otherwise, the effects of datura seemed more amusing than dangerous.[74] In the 1880s, A. H. Giles (who sought to downplay the notoriety attached to Indian poisons) referred to its use 'by way of frolic'. He cited an occasion when 'the whole of a jail guard in the North-Western Province [sic] was ... placed hors de combat on the floor of their guard-room', but sadly gave no details.[75] In its medicinal guise, datura was used by Europeans in India as an antispasmodic: its leaves were smoked as a treatment for asthma, and, for a while, as one of many East-West drug exchanges, the 'divine stramonium' was put to similar uses in Europe before falling out of favour in the late nineteenth century.[76] There were, however, more sinister references to datura as a 'pernicious' drug and an 'intoxicating poison' dating back to 1816, even if it was rarely seen as life threatening – more the kind of substance with which 'wicked boys' might entertain themselves by adding datura to the food served at a wedding feast.[77] In 1841, O'Shaughnessy noted the use of datura in Bengal to facilitate theft and 'other criminal designs', but felt sure this was done 'to stupefy merely' and 'not with the intention of killing'. In his experience, even 'intoxication or delirium' was 'seldom produced'. The victim sank into a 'profound lethargy, resembling coma, with dilated pupils, but natural respiration'. Recovery normally occurred within two days. O'Shaughnessy added, however, that, on an empty stomach, a small quantity could cause a more extreme reaction and even prove fatal. Since this fact was 'well known to the Indian poisoners', they suited the size and timing of their dose 'according to the purpose they mean to serve'.[78]

---

[73] S. P. Sangar, 'Intoxicants in Ancient India', *Indian Journal for History of Science* 16 (1981): 204–14; Kanny Lall Dey, *The Indigenous Drugs of India* (2nd ed., Calcutta: Thacker, Spink, 1896), 111–12.

[74] Garcia da Orta, *Colloquies on the Simples and Drugs of India* (trans. Clement Markham, London: Henry Sotheran, 1913), 174–77.

[75] Giles, 'Poisoners', 98.

[76] G. Skipton, 'Three Cases Showing the Beneficial Effects of Dhatura in Asthma', *TMPSC* 1 (1825): 121–23; Mark Jackson, '"Divine Stramonium": The Rise and Falling of Smoking for Asthma', *Medical History* 54 (2010): 171–94.

[77] Judicial letter to Bombay, 10 April 1815; Judicial letter from Bombay, 6 August 1816; Bombay, Judicial Consultations, 29 May 1816, F/4/545: 13314, IOR.

[78] O'Shaughnessy, *Bengal Dispensatory*, 469.

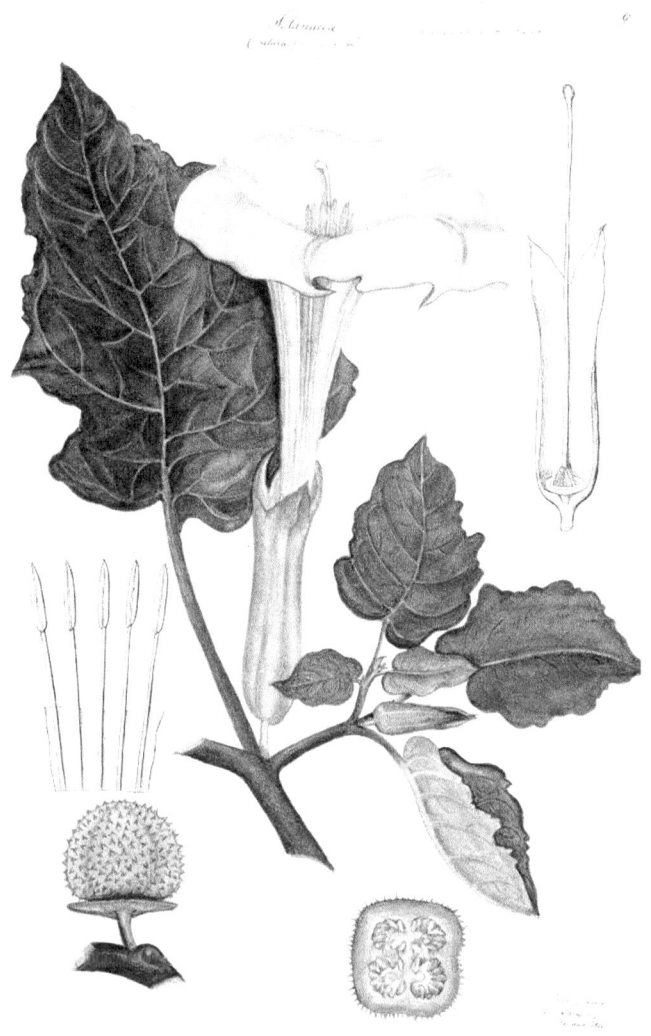

Figure 6. Datura fastuosa from Hugh Cleghorn's collection of Indian botanical illustrations. The caption (bottom left) reads 'Springs up on rubbish, and seems to be one of those plants which follow man', and (bottom right) 'Very common. Shimoga [Mysore], 29th Augt. 1845'.

As the criminal associations of datura grew during the nineteenth century, colonial medicine became increasingly sceptical about its safe use and medical efficacy. India's surgeon-botanists were not just giving a professional verdict on the properties of datura; they were also, as servants of the Raj, thinking collectively, and 'like a state', about its criminal dangers and the threat it might pose to public health and public order. In 1868, Edward Waring described datura as effective in relieving pain in rheumatism and for treating boils and tumours, but he hastened to add: 'Its poisonous properties in large doses are well known to the natives [of India], who employ it frequently for criminal purposes.' No evidence was given to support the incriminating use of the word 'frequently'.[79] Balfour, who devoted an entry in his *Cyclopaedia* to datura, commented that its 'value as a curative in asthma is known both to Europeans and natives, who smoke the seed in their hookahs when so afflicted'. But he, too, warned that thieves used datura to stupefy innocent travellers. 'The victims are usually discovered in a state of insensibility and breathing hard and heavily; if removed, care should be taken not to expose them to the heat of the sun, which is fatal.'[80]

Articles appeared in India's medical press highlighting the criminal use of datura and claiming that many of its victims 'only narrowly escaped death'.[81] Witnesses to the Indian Hemp Drugs Commission explicitly referred to datura as a poison, whose effects when used by 'professional poisoners' were far worse than those of mere intoxicants like ganja.[82] In a trenchant account of Indian criminality in 1909, H. L. Adam devoted several chapters of his book to the guile of 'the Eastern poisoner'. 'One of the worst forms of poisoning', datura figured prominently in this alarmist narrative. Its effects were said to be 'so disastrous that death itself ... would come as a merciful relief'.[83] Adam continued:

It impairs their [the victims'] faculties, and sometimes permanently deranges their minds. In very few cases have the victims been known to regain their former bodily and mental vigour, and then only where but a light dose of the poison happens to have been administered. One man who had been dosed still remained a cripple seven years after the event; another victim was scarcely able to articulate, as

---

[79] Waring, *Pharmacopoeia*, 175.

[80] Balfour, *Cyclopaedia*, 1: 897. Cf. W. J. Moore, *A Manual of Family Medicine for India* (4th ed., London: J. & A. Churchill, 1883), 558–59.

[81] David R. Smith, 'Suspected Criminal Poisoning by Dhatoora in the Person of a European', *IMG* 3 (1868): 58–60; J. A. Greene, 'Case of Poisoning by Dhatoora', *IMG* 6 (1871): 165; J. Cleghorn, 'Cases of Datura Poisoning', *IMG* 6 (1871): 209.

[82] *Indian Hemp Drugs Commission, Vol. 5: Evidence of Witnesses from North-Western Provinces and Oudh and Punjab* (Calcutta: Office of the Superintendent of Government Printing, India, 1894), 52, 109, 119, 161, 269, 336.

[83] H. L. Adam, *The Indian Criminal* (London: John Milne, 1909), 100.

though he had been struck by paralysis; he had never left his bed, and was gradually wasting away.[84]

We will see in later chapters how evidence for datura's criminal use was assembled. But it is worth stating here that, even though occasional cases of robbery by means of datura were reported into the 1920s and 1930s, very few cases of death or serious injury were recorded. In 1870–72, datura accounted for only 0.27 per cent of all reported poisonings in northern and eastern India (83 out of 31,310).[85] In 1911, in the United Provinces and Central Provinces, datura was blamed for only two deaths, 0.5 per cent of all poison fatalities in that year.[86] Such meagre statistics might perhaps indicate difficulty in detecting datura poisoning, especially since datura seeds (the main toxic agent) were easily confused with those of the common capsicum (much as arsenic poisoning might be confused with cholera).[87] But the apparent paucity of fatal datura poison cases more strongly suggests that the extreme claims made for the drug's toxicity were greatly exaggerated. By the 1960s, the old association of datura with thugs and highway robbers was still distantly recalled, but there was general scepticism about its allegedly lethal properties. Chopra and his associates concluded that a fatal outcome was rare, and, mixed with bulky foods like rice, its effect was mild, with recovery within days. Its use in murder was uncommon and in suicide exceptional.[88] Datura had by then shed most of its demons.

Opium wars

Of all the many drugs Western medicine shared with its indigenous counterparts, opium was much the most common. Even after the introduction of quinine as a prophylactic and treatment for malaria, opium remained India's most popular medicinal drug. But opium was not a substance known to ancient medical authors: it only gained widespread acceptance in India after about the twelfth century.[89] It then became extensively used as a febrifuge, analgesic, aphrodisiac and remedy for

[84] Ibid., 100–01.
[85] Robert Harvey, 'Report on the Medico-Legal Returns Received from the Civil Surgeons in the Bengal Presidency during the Years 1870, 1871, and 1872', *IMG* 11 (1876): 60.
[86] *ARCE (UP), 1911*, 5.
[87] On the distinction between datura and capsicum seeds, see Alexander Wynter Blyth and Meredith Wynter Blyth, *Poisons: Their Effects and Detection* (4th ed., London: Charles Griffin, 1906), 379.
[88] Chopra et al., *Poisonous Plants* 2: 638, 645–46.
[89] Udoy Chand Dutt, *The Materia Medica of the Hindus* (Calcutta: Thacker, Spink, 1877), xi; G. N. Chaturvedi, S. K. Tiwari and N. P. Rai, 'Medicinal Use of Opium and Cannabis in Medieval India', *Indian Journal of History of Science* 16 (1981): 31–32.

dysentery and diarrhoea: it was even recommended as an antidote for datura.[90] Widely grown across western, central and northern India, opium became integral to the ceremonial rites and social life of the Rajput warrior caste, for whom it was said to be 'more necessary than food'.[91] Not all commentators regarded this reliance on opium as beneficial or even benign. In 1879, G. R. Aberigh-Mackay, Principal of Rajkumar College in central India, regretted the decline of the Rajput chiefs (whose sons were among his charges) from their former martial bearing to a state of 'slothful ignorance and debauchery'. He attributed their 'dull, morose and wretched condition' not to their political eclipse by the British, which might have been one explanation, but to the deleterious effects of the drug.[92]

If British medical practitioners were acquainted with opium's many positive attributes before they reached India, they were also likely to be familiar with reports in the medical press about accidental opium poisoning, especially among children.[93] However, in India, they tended to adopt a more positive view of its therapeutic utility, and favourable – at times euphoric – references to the medicinal use of opium abound in nineteenth-century texts. When epidemic cholera erupted in India in 1817, opium pills were prescribed as one of the few means of controlling so rapid and deadly a disease. In the days when physicians still favoured bloodletting in fever cases, opium was used, alongside calomel, to restore the patient and combat nervous exhaustion. Its use as a painkiller and soporific even extended to the treatment of smallpox.[94] Opium had uses that ranged well beyond the medicinal. Administered as a small pill or mixed with spices to form tiny pellets, opium was thrust into the mouths of infants to make them docile. In the early twentieth century, this practice was apparently routine among women workers in the Bombay textile mills, though when Indian ayahs did the same with their European charges, it caused a degree of alarm.[95] And, as we have already seen,

[90] In the 1930s, Chopra denied that opium *was* widely used by *vaid*s and hakims, but this may have reflected recent decline in its therapeutic use: R. N. Chopra, *Indigenous Drugs of India: Their Medical and Economic Aspects* (Calcutta: N. Mukherjee, 1933), 202.

[91] James Tod, *Annals and Antiquities of Rajasthan* (3 vols, London: Oxford University Press, 1920), 2: 750.

[92] G. R. Aberigh-Mackay, *The Chiefs of Central India* (Calcutta: Thacker, Spink, 1879), xxx–xxxi, lxxvi.

[93] Christison, *Treatise*, 530–71.

[94] W. L. McGregor to Superintendent of Hill States, Simla, 30 April 1848, F/4/2401: 129687, IOR; O'Shaughnessy, *Bengal Pharmacopoeia*, 399; James Ranald Martin, *The Influence of Tropical Climates on European Constitutions* (London: John Churchill, 1856), 168.

[95] F. D. Barnes, 'Problems Relating to Working Mothers and Infants', *Social Service Quarterly* 8 (1922): 14; Moore, *Manual*, 558.

opium was implicated in female infanticide. It may also have been used to pacify widows as they were led to the funeral pyre in the rites of sati.[96]

Opium exemplified the dilemmas of the *pharmakon*. Used in small 'medicinal' doses, it was hailed as an invaluable therapeutic; given in excess (beyond ten or a dozen grains a day), it became a dangerous 'poison'. Its medicinal use was further complicated by the different varieties of opium – Smyrna, Malwa, Patna and 'hill opium' from the Himalayan foothills – and the different percentages of morphia they contained.[97] Defenders of opium argued that it was invaluable, too, because its properties were so widely known to the Indian population, and *vaid*s and hakims could be trusted to prescribe its safe use. That opium *could be* a poison was not denied, but the same could be said of many items of food and drink, including the vast quantities of hard liquor swilled down by European soldiers in India. Taking opium was arguably no worse than drinking beer (and certainly arrack, which was sometimes fortified with datura) or smoking tobacco, and it was thought to pose less danger to health than ganja.[98] This relatively benign view of opium appears to have been the dominant one among most European doctors in India and among the growing number of Indian practitioners of Western (or allopathic) medicine.[99] But there always were physicians wary of the medicinal use of opium and some who shunned it entirely. In 1880, in discussing the effects of opium addiction among prisoners, an Indian assistant surgeon stated, 'opium is a poison'.[100] A decade later, Waring warned that, even in the treatment of cholera, its use in large and repeated doses was 'fraught with danger'.[101]

By the mid-nineteenth century, opinions about opium could not be detached from the controversy raging over the drug generally, especially the huge volume of opium sold in India or exported to China.[102] Unlike datura or *Aconitum*, which were of minor financial significance, opium was big business: in 1871–72 alone, it netted the government close to

---

[96] [Marianne] Postans, *Cutch; or, Random Sketches, Taken during a Residence in One of the Northern Provinces of Western India* (London: Smith, Elder, 1839), 65. The same narrative relates the use of opium in female infanticide: ibid., 145–46.

[97] McGregor to Superintendent of Hill States, 30 April 1848, F/4/2401: 129687, IOR.

[98] Editorial, 'The Crusade against Opium', *IMG* 27 (1892): 178–80; W. J. Moore, *The Other Side of the Opium Question* (London: J. & A. Churchill, 1882); Patrick Hehir, *Opium: Its Physical, Moral, and Social Effects* (London: Ballière, Tindall & Cox, 1894).

[99] Editorial, 'The Opium Commission', *IMG* 29 (1894): 21.

[100] Ram Chunder Mitter, letter, *IMG* 15 (1880): 223.

[101] Edward John Waring, *Remarks on the Uses of Some of the Bazaar Medicines and Common Medicinal Plants of India* (5th ed., London: J. & A. Churchill, 1897), 116.

[102] Carl A. Trocki, *Opium, Empire and the Global Political Economy: A Study of the Asian Opium Trade, 1750–1950* (London: Routledge, 1999); Hans Derks, *History of the Opium Problem: The Assault on the East, ca. 1600–1950* (Leiden: Brill, 2012).

£8 million in revenue.[103] In the critics' view, the opium trade was a crime, just as surely as opium was an addictive, life-sapping poison. Pressure from the anti-opium lobby forced the British government to accede to demands for a royal commission to investigate opium production and sale, but when the commission's report appeared in 1895, it was widely seen as exonerating the Government of India and ignoring opium's critics.[104] The report has been regarded by some scholars, too, as a blatant example of how medical opinion in India was mobilized to uphold the regime's dubious economic and political interests, not least through the spurious claim that opium had great value in the treatment of malaria, among India's most deadly diseases. The report appeared to condone Britain's profitable but highly immoral opium trade.[105] But the position of physicians in India was more complex than this argument suggests. As we have already seen, Western medicine in India could be highly critical of indigenous drugs and their toxic properties, but the medical profession also had a long history of absorbing local therapeutic ideas and practices. Since few Ayurvedic and Unani physicians regarded opium as dangerous, let alone poisonous, it was hardly surprising that a relaxed view of its toxicity prevailed among most Western doctors.

Because opium had recreational as well as medicinal uses, it was possible for physicians to cite examples of high levels of tolerance among regular consumers of the drug while at the same time arguing, especially from prison experience, that opium was not seriously addictive and its denial did not cause unmanageable craving.[106] While critics argued that the cultivation of opium poppies had helped cause famine in Orissa in 1866 and Rajasthan in 1878 by diverting peasants from growing food crops, others, like Surgeon-General W. J. Moore, responded by saying that opium consumption had 'proved the salvation of many' by enabling them to 'bear up against hardships which they could not otherwise have withstood'.[107] If opium acted as a poison, it was because individuals chose to use it that way (as to commit suicide): but

---

[103] Martin Booth, *Opium: A History* (New York: Simon & Schuster, 1996), 8–10.
[104] David Edward Owen, *British Opium Policy in China and India* (New Haven: Yale University Press, 1934), ch. 11. For a defence of the report, see John F. Richards, 'Opium and the British Empire: The Royal Commission of 1895', *Modern Asian Studies* 36 (2002): 375–420.
[105] Paul C. Winther, *Anglo-European Science and the Rhetoric of Empire: Malaria, Opium, and British Rule in India, 1756–1895* (Lanham: Lexington Books, 2003). Cf. William Roberts, 'The Opium Habit in India', in *Collected Contributions on Digestion and Diet* (2nd ed., London: Smith, Elder, 1897), 289–98. Chopra later stated that opium had 'neither a prophylactic nor a curative action' on malaria: Chopra, *Indigenous Drugs*, 214.
[106] The prison was an important site for the observation of the effects of opium and its withdrawal: see Thomas Mayne, 'Opium-Eaters', *IMG* 16 (1881): 89–90.
[107] W. J. Moore, 'The Opium Question', *IMG* 16 (1881): 213.

to try to ban a substance so widely used and socially acceptable was unthinkable, especially after the uprising of 1857–58 had made the British chary about meddling with 'native customs'. Patrick Hehir observed in 1894 that, even if a prohibition were introduced, it could never be enforced: 'There are so many difficulties in the way not met [with] in England and Europe ...'[108] Moreover, the default claim that 'India was different' could be used to distance opium use in India from addiction in China. If the Chinese succumbed in their millions to opium, then that, the drug's defenders claimed, demonstrated a psychological or racial weakness not to be found among Indians.[109]

This, then, was the reverse of datura. In defiance of Indian cultural values and therapeutic traditions, datura was demonized as dangerous. With opium, by contrast, what many in the West regarded as a poison, whose very exoticism lent credence to its reputation for toxicity, in India was commended as a familiar drug and medicinal standby. Poisons were substances that killed or caused serious injury; they aided the commission of crime. Datura fitted both these criteria in ways that a mere intoxicant (like cannabis) seemingly did not. Certainly, as what some saw as a 'narcotic poison', opium fell into a more ambiguous category.[110] But in India opium had few criminal associations and (except in suicides) possessed an apparently greater capacity to cure disease or quell pain than to occasion serious loss of life. While datura was an ungovernable weed that sprang up almost anywhere, opium was a highly domesticated drug. Its sale was already regulated by the state, and it was, of course, a great financial asset to the British government in India. Opium's immense commercial value was, no doubt, responsible in part for the very different assessment of opium's poison status compared to datura. But Indian cultural values and therapeutic practices, and their impact on colonial thinking, cannot be ignored.

Toxic exchanges

In recent scholarship on the history of medicine in British India, there has been extensive discussion about the relationship between 'indigenous' and 'colonial' medicine. Some scholars have argued that there was an ongoing dialogue between them, a two-way traffic, perhaps less marked after the 1830s than previously, but continuing throughout the nineteenth century and beyond. Others have taken the opposing view that even in the early nineteenth century, allopathic physicians looked upon

---

[108] Hehir, *Opium*, 357.   [109] Moore, *Other Side*, 36.
[110] Beck and Beck, *Elements*, 865–82.

indigenous medicine and its practitioners with scepticism and scorn, and, as the century progressed, with outright hostility.[111] Since no medical system was entirely homogeneous and the context – of location, disease, type of practitioner – varied widely, perhaps no comprehensive assessment can be made. But the subject of poisons has a particular bearing on this debate. Because of the extent of India's pre-colonial poison knowledge and the abundance of substances in India classed as poisons and/or medicines, Indian texts, Indian experience and the voice of Indian practitioners commanded real authority, even among British physicians in India who might otherwise have appeared cautious or censorious. Here was a domain of scientific knowledge and medical practice where Indians were often better informed than Europeans.

It is striking how frequently Indian knowledge and Indian use of toxic substances were cited in colonial sources. It is, for instance, suggestive of a continuing dialogue that as late as the 1890s Dymock included in his *Pharmacographia* the views of Indian physicians and even texts composed centuries earlier. Thus, in his account of *Aconitum ferox*, he noted that the *Makhzan al-Adwiyah*, a famed compendium of Unani drugs, had identified *bish* as an Indian root with 'hot' and 'dry' properties, suited for use internally in the treatment of leprosy, asthma, coughs and ulceration of the throat, and externally against boils and neuralgia.[112] In discussing datura, Dymock again cited the *Makhzan al-Adwiyah* for its vivid description of the drug's effects:

Everything he [the patient] looks at appears dark; he fancies that he really sees all the absurd impressions of his brain, his senses are deranged, he talks in a wild, disconnected manner, tries to walk but is unable, cannot sit straight, insects and reptiles float before his eyes, he tries to seize them, and laughs inordinately at his failure. His eyes are bloodshot, he sees with difficulty, and catches at his clothes, and the furniture and walls of the room. In short, he has the appearance of a mad man.[113]

Can we interpret this as evidence of a genuine and reciprocal dialogue between East and West? Certainly, Dymock, like many other European authors, saw value in the observations made by non-Western physicians. Indeed, the accounts he cited of the physiological and psychological effects of datura from indigenous texts were seldom bettered by European writers. In consulting Indian expertise, through texts or

---

[111] For recent comments on this debate, see Ishita Pande, *Medicine, Race and Liberalism in British Bengal: Symptoms of Empire* (Abingdon: Routledge, 2009); Amna Khalid and Ryan Johnson, 'Introduction', in Ryan Johnson and Amna Khalid (eds), *Public Health in the British Empire: Intermediaries, Subordinates, and the Practice of Public Health, 1850–1960* (New York: Routledge, 2013), 1–31.

[112] Dymock, *Pharmacographia*, 1: 2.   [113] Ibid., 2: 587.

conversational exchanges, there remained the possibility, too, that Indian pharmacology contained information on unknown or neglected drugs that might profitably be incorporated into the allopathic system. But Dymock also sought to distance himself from the imperfect knowledge, as he saw it, of Indian *vaids* and hakims, who depicted datura as merely an intoxicant and not as the dangerous drug and instrument of criminal poisoning that British experience had revealed it as.[114] Similarly, while Waring paid tribute to what he had learned from 'Native sources', he also claimed the credit for repaying that debt with interest by adding 'a considerable amount of information on the uses of even the same drugs, of which the Natives themselves had previously no idea'.[115]

Even apart from their rejection of humoral diagnostics and pathology, there was a further reason for Western doctors to take indigenous pharmacology seriously without necessarily subscribing to its precepts and practices. As Udoy Chand Dutt observed in 1877, the allopathic physician needed to know the nature and composition of indigenous medicines 'when he comes to hear of their having been used by patients who had been under native treatment before coming under his care, as is very often the case'.[116] As healers of 'last resort', practitioners of Western medicine had to know about the 'poisons' their Ayurvedic and Unani counterparts had used on those who later became their own patients.[117] It was often in this negative context, rather than from positive appreciation, that Western medical texts referred to aphrodisiacs and abortifacients. It became increasingly common for practitioners of Western medicine to identify hakims, *vaids* and the *pansaris* who supplied their drugs as, in effect, peddling poisons and, from ignorance or ill will, prescribing substances that killed or maimed their patients. Part of the self-legitimation of Western medicine lay in this denunciation of the inferiority of indigenous medicine and its criminal or murderous associations. Calling local drugs and medicinal compounds 'poisons' provided a convenient means of demarcating what was seen to be legitimate practice from improper or 'quack' medicine. An article in the *Indian Medical Gazette* in 1867 strongly denounced both hakims and *pansaris*. The 'unlicensed practice of Native Hakeems' and 'the unrestricted sale of poisonous drugs' by the *pansaris* were, the author remarked, 'fraught with dire consequences ... Life is daily jeopardised, and the door to crime and knavery thrown wide open.' Forty years before a poisons act and medical registration acts came into being, the article urged the government to end this 'barbarous state of

---

[114] Ibid., 2: 590.   [115] Waring, *Remarks*, xiii.   [116] Dutt, *Materia Medica*, iv.
[117] Alan R. Beals, 'Strategies of Resort to Curers in South India', in Charles Leslie (ed.), *Asian Medical Systems: A Comparative Study* (Berkeley: University of California Press, 1976), 184–200.

things'.[118] Not all of the *Gazette*'s correspondents and contributors took quite such a hostile view of the 'murderous' hakim (*pansari*s mustered few defenders).[119] But the argument that the indigenous systems were not only erroneous but deadly continued to gain ground, reinforced by the publication of cases in which *vaid*s and hakims had prescribed aphrodisiacs, anti-worm powders or cures for syphilis that turned out to contain fatal doses of datura, arsenic or oleander.[120]

And yet a dialogue was maintained, and Indians, trained in the Western medical system, were largely instrumental in this knowledge exchange. Even the most censorious of British physicians in India, such as Norman Chevers (see Chapter 4), made appreciative reference in their published work to Indians who had supplied them with information about poisonous drugs and poison practices. In particular, Kanny Lall Dey and Udoy Chand Dutt in the 1860s and 1870s and Chunilal Bose in the 1890s and 1900s played a crucial role in bringing India's poison drugs and a scientific knowledge of their uses and properties to Western attention. There was nothing apologetic about this. In 1862, Dey compiled a list of Indian drugs sent to London for an exhibition that included poisons and abortifacients. In 1867, he published a more extended account of *The Indigenous Drugs of India*, which incorporated medicines and poisons sold in bazaars or used by Bengali *vaid*s and hakims.[121] In his own research, he devised an improved method of testing for the presence of Indian opium in forensic samples.[122] In 1877, Dutt published a *Materia Medica of the Hindus*, compiled from Sanskrit medical works, in which poisons again enjoyed prominence.[123] And, in a series of scientific papers in the 1890s and 1900s, Chunilal Bose, as Bengal's Chemical Examiner, presented research on medicinal plants and on narcotic and toxic substances from *bish* (and 'false *bish*') to oleander root and cocaine.[124]

---

[118] Extract from *Indian Public Opinion*, in *IMG* 2 (1867): 23.

[119] For *pansari*s as poisoners, see Mortimer Menpes and Flora Annie Steel, *India* (London: Adam & Charles Black, 1905), 93. Cf. Editorial, 'A Plea for Hakeems', *IMG* 3 (1868): 87–89.

[120] T. Hume, 'Case of Partial Paralysis, Supposed to Have Followed the Injudicious Administration of Arsenic', *IMG* 11 (1876): 103; Chunilal Bose, 'On the Chemistry and Toxicology of *Nerium odorum*', *IMG* 36 (1901): 287–90.

[121] Kanny Lall Dey, 'Medicinal Substances Used by Native Practitioners', in A. M. Dowleans, *Official Classified and Descriptive Catalogue of the Contributions from India to the London Exhibition of 1862* (Calcutta: Bengal Printing Co., 1862), 65–81; idem, *The Indigenous Drugs of India*.

[122] F. N. Macnamara, 'Report of the Chemical Examiner to Government for the Year Ending March 1874', Bengal, Medical, 172-12/13, 27 May 1874, IOR.

[123] Dutt, *Materia Medica*.

[124] See J. P. Bose (ed.), *The Scientific and Other Papers of Rai Chunilal Bose Bahadur* (2 vols, Calcutta: Forward Press, 1924).

All three – Dey, Dutt and Bose – acted as more than just willing intermediaries between two very different knowledge systems. They were also advocates and educators, seeking to inform British practitioners of Western medicine and pharmacology about the nature and potentialities of Indian medicines and poisons while seeking, too, to educate their own class and country about India's rich medical and toxicological inheritance. And it was not only these three, pre-eminent though their contribution was. Numerous articles and communications from Indians in the subordinate medical services were published in the *Indian Medical Gazette*, the leading medical journal at the time.[125] These drew upon the authors' first-hand knowledge of poison cases in hospitals, prisons and dispensaries and demonstrated their superior knowledge of the social and cultural milieu of the Indian people whose encounters with toxicity – in food, drink, medicaments and 'love potions' – they described. Indians were not only better informed by dint of their greater access to Sanskrit texts, vernacular sources and personal contact with *vaid*s and hakims; they were also closer than most Europeans to the quotidian experience of health, disease and poisons among the Indian populace.

The second half of the nineteenth century saw a momentous 'revitalization' of Ayurvedic medicine.[126] This movement drew strength from the arrogance of Western medicine and its repeated denigration of Indian medical knowledge, but it was also grounded in a conviction, especially among the *kaviraj*s, the Ayurvedic practitioners and ideologues of Bengal, that India's medical systems, tried and tested over millennia, retained vast reserves of un-utilized medical knowledge and physiological understanding. One way of seeking to demonstrate that the indigenous systems were no less efficacious or scientific than their Western counterpart, and to rebut repeated criticism levied against them of ignorance and recklessness, was to emphasize India's rich store of poison knowledge and show the careful and constructive medicinal use of such potent substances.[127] Far from disavowing India's poison culture, these texts boldly reaffirmed the centrality of

---

[125] For example, Tarra Prosonno Roy, 'On the Use of Dhatura as a Mydriatic', *IMG* 5 (1870): 187–88; Hem Chunder Bhuttacharjee, 'Case of Poisoning by *Gloriosa superba*', *IMG* 7 (1872): 153.

[126] K. N. Panikkar, 'Indigenous Medicine and Cultural Hegemony: A Study of the Revitalization Movement in Keralam', *Studies in History* 8 (1992): 283–308; Charles Leslie, 'The Ambiguities of Medical Revivalism in Modern India', in Leslie (ed.), *Asian Medical Systems*, 356–67; Uma Ganesan, 'Medicine and Modernity: The Ayurvedic Revival Movement in India, 1885–1947', *Studies on Asia* 4 (2010): 108–31.

[127] Nagendra Nath Sen Gupta, *The Ayurvedic System of Medicine* (2 vols, Calcutta: Keval Ram Chatterjee, 1901), 1: 474; Gananath Sen in *The Report of the Committee on the Indigenous Systems of Medicine* (2 vols, Madras: Government Press, 1923), 2: 1–6.

toxicology within Ayurveda and suggested that it was the very use of such powerful drugs that gave many indigenous cures their potency.[128] This vigorous defence worked in tandem with the view that indigenous therapeutics were better suited to India's 'tropical' environment, to 'the children of the soil' and to Indians' cultural needs and physical make-up than 'the administration of violent foreign medicines [which] so affects the debilitated systems of [Indian] patients'.[129] It was only a small step from this position to turning the tables on Western medicine entirely by describing it (rather than Ayurveda or Unani medicine) as a purveyor of 'deadly poisons'.[130] The debate over which was the more poisonous persisted into the 1930s and 1940s: indeed, it has never quite disappeared.[131]

All systems of medicine navigate the perils of the *pharmakon*, to decide how to differentiate between medicines that heal and poisons that kill and in what circumstances and proportions. This was certainly the case in nineteenth-century India, but the task there was complicated by the extent of India's pre-existing knowledge and use of poisons and by a political situation in which toxicity became the compelling vehicle for a highly negative view of India and its indigenous medical systems. Within the space created by empire, there existed a dual process of engagement. On the one side, India's allopathic medicine was in active communication with metropolitan Europe. O'Shaughnessy brought forensic chemistry to India and established the role of the chemical examiner in colonial poison governance. Conversely, Royle warned against misuse of dangerous Indian drugs like *Aconitum ferox* but also, capitalizing on his India expertise, promoted their therapeutic use in the West. On the other side of the imperial equation, allopathic medicine was in earnest (if at times one-sided) conversation with *vaid*s and hakims, with Indians like Dey and Dutt, trained in Western medicine, who nurtured an awareness of indigenous toxicological knowledge while also establishing an authority of their own – for themselves and for their science – by re-presenting and reinterpreting this knowledge. The imperial *pharmakon* – the space opened up across

[128] B. D. Basu, 'On the Study of Indigenous Drugs', *IMG* 28 (1893): 336–38.

[129] Binod Lall Sen and Athutosh Sen, 'Preface', to Udoy Chand Dutt, *The Materia Medica of the Hindus* (2nd ed., 1900, reprinted Delhi: Mittal Publications, 1989), iii; Jamini Bhushan Roy, in *Report of the Committee on the Indigenous Systems*, 2: 14–15.

[130] John Martin Honigberger, *Thirty-Five Years in the East* (London: H. Ballière, 1852), xiii; G. Srinivasa Murti, 'A Memorandum on the Science of the Art of Indian Medicine', in *Report of the Committee on the Indigenous Systems* 1, appendix 1: 96.

[131] *Report of the Drugs Enquiry Committee, 1930–31* (Calcutta: Government of India, Central Publications Branch, 1931), 354–67.

the empire for therapeutic exchanges and toxic dialogue – was not solely a British creation or concern. The conditions that made the emergence of the science of toxicology possible in India owed much to the scientific interests, political imperatives and metropolitan engagements of the colonial regime. But Indian agency and Indian knowledge of poisons and their local usages also constituted an essential part of that process.

3   Panics and scares

For much of the nineteenth century, poisons lived quiet lives, seldom exposed to public scrutiny or state sanction, slipping inconspicuously in and out of Indians' daily existence. Their presence – if we think of murder or suicide or accidental death – might be tragic enough in their immediate effects and personal consequences, but poison in general was unlikely to shatter the peace of a society already beset by famines, epidemics, mass poverty and periodic revolt. Nor was poison likely to subvert the wider political and social order: rather, the acts of suicide, infanticide and foeticide in which opium, aconite, arsenic and *lal chitra* were implicated might even serve to stifle dissent and uphold social norms. And yet from time to time and in certain contexts, poisons – and the anxieties they occasioned – took on a more menacing character, threatening enough to foster social unease and an implicit questioning of state authority. Sometimes, these eruptions, these toxic fears, were temporary phenomena and died away as quickly as they had appeared. At other times, they were not so easily dissipated or resolved.

Poison might emerge from these moments of crisis less as a kaleidoscope of wayward social, therapeutic and criminal practices and more as a concept, elevated to a public meaning and political significance beyond individual action and personal mishap. Poison, or – as we move in this chapter from substances to social entities – those suspected of being poison's purveyors, might thereby call for a greater level of scientific engagement by the state and from the public. This chapter looks, therefore, at the way in which poison crises or 'panics' arose, examines their consequences in the light of what has been termed the colonial 'information order' and shows how they impelled a new scientific and administrative approach to poisons and poisoning by the late nineteenth century.

Poison panics

Panic was no stranger to India. Under British rule, the country experienced a series of intermittent scares, some escalating beyond nervous rumour and incipient alarm into wholesale, full-blown panics. Some of these episodes arose from fear of insurrection or invasion, and these were by no means confined to the Indian population. Calcutta was thrown into turmoil during 'panic Sunday' on 14 June 1857 as white residents feared that the sepoys (soldiers) who had begun to mutiny the month before in northern India were about to descend on the city and slaughter them, or that its restive Muslim population was going to rise up in arms against them.[1] Other panics, more typically among Indians, were precipitated by sudden food shortages or arose from alarm at the rapid spread of deadly epidemics of cholera and plague.[2] Panics and the rumours that fuelled them were thus a relatively common phenomenon in nineteenth- and early-twentieth-century India, and they can be seen as demonstrating the inherent weakness of what C. A. Bayly termed the colonial 'information order' – the unsettling inadequacy of British knowledge and understanding of what was actually happening in India, or being communicated, in a fervour of rumour and report, among Indians of all classes.[3]

There was, however, a type of rumour and panic in India that has not hitherto received due comment, one that played upon fears of poisoning or used the idiom of poison to articulate a wider mood of collective unease.[4] Poison panics arose less from fear of external assault, from an invading force or armed insurrection, than from fear of subversion from within (within society, within the human body) by means of impure or contaminated food and water. Poisoning and the fear it occasioned is a phenomenon noted for a number of slave-holding societies around the globe and as such has often been seen as a covert form of retribution and insurrection. In her discussion of creole society in eighteenth-century Mauritius, Megan Vaughan describes poisoning as being 'in some ways the colonial crime par excellence', adding that 'tales of poisoning abound from all over the colonial world'. Poisoning, she writes, was 'a crime of stealth', an act 'of the powerless'. She suggests that for plantation slaves, like those on sugar estates in Mauritius, it could be a means by which,

---

[1] [John Kaye and G. B. Malleson], *Kaye's and Malleson's History of the Indian Mutiny of 1857–8* (6 vols, London: W. H. Allen, 1889–93), 4: 19. For earlier 'white panics', see Peter Robb, *Sentiment and Self: Richard Blechynden's Calcutta Diaries, 1791–1822* (New Delhi: Oxford University Press, 2011), 34.

[2] David Arnold, *Colonizing the Body: State Medicine and Epidemic Disease in Nineteenth-Century India* (Berkeley: University of California Press, 1993), chs 4–5.

[3] A. A. Yang, 'A Conversation of Rumors: The Language of Popular *Mentalités* in Late Nineteenth-Century Colonial India', *Journal of Social History* 20 (1987): 485–505; C. A. Bayly, *Empire and Information: Intelligence Gathering and Social Communication in India, 1780–1870* (Cambridge: Cambridge University Press, 1996).

[4] The idea of 'poison panics' is, though, well established in the British context: e.g. Ian Burney, *Poison, Detection, and the Victorian Imagination* (Manchester: Manchester University Press, 2006), 20.

without the hardships and hazards of outright revolt, the oppressed could exploit the physical vulnerability of their masters, spreading alarm among the slave-holding class without forfeiting their own safety and anonymity.[5]

Poison scares and panics in British India were seldom so racially exclusive or so evidently the province of the 'powerless'. In some instances, these 'moral panics' (to use Stanley Cohen's now somewhat hackneyed phrase) might reflect the anxieties of only one section of colonial society, European or Indian, but others crossed racial and class divisions and, by revealing shared beliefs in the 'folk devils' responsible, helped inspire a wider sense of public danger.[6] It is worth noting, too, in the light of Cohen's discussion, that in the early and middle decades of the nineteenth century, newspapers in India were still in their infancy, and so the mass media that figured so prominently in his account of mods and rockers in 1960s Britain played little part in the dissemination of toxic tales and their demonic imagery. Instead, oral transmission by means of rumour, especially the 'bazaar *gup*' of which European observers were both contemptuous and wary, was far more important in generating alarm, just as the anxieties of the colonial bureaucracy and white ruling class largely circulated through the restricted Anglophone medium of official correspondence.[7] Like the concern raised by F. J. Mouat as Bengal's Chemical Examiner in 1843 (see Chapter 2) over poisonous drugs on sale in the bazaar, many of these panics focused on the market-place, for most Indians (and many Europeans) a critical point of contact and exchange, a 'space of experience' where contamination, pollution and poison were ever-present possibilities.[8]

What were these 'poison panics'? The idea of panic-causing poisoning conveyed a range of cultural and corporeal meanings, linking toxicity, real or imagined, to items of everyday consumption, religious identity and social status. In the mid-1830s, salt sold in the markets of Saran district in Bihar was said to contain sediment resembling the ground-up bones of cattle, animals sacred to Hindus. Although dismissed by Robert Rankine, the civil surgeon, as a 'ridiculous story' and 'absurd report', the rumour was sufficiently disturbing to provoke a boycott of traders by Hindus

---

[5] Megan Vaughan, *Creating the Creole Island: Slavery in Eighteenth-Century Mauritius* (Durham, NC: Duke University Press, 2005), 98.

[6] S. Cohen, *Folk Devils and Moral Panics: The Creation of the Mods and Rockers* (3rd ed., London: Routledge, 2002); Kenneth Thompson, *Moral Panics* (London: Routledge, 1998).

[7] John Campbell Oman, *Cults, Customs and Superstitions of India* (London: T. Fisher Unwin, 1908), 218–28.

[8] Reinhart Koselleck, *Futures Past: On the Semantics of Historical Time* (Cambridge, MA: MIT Press, 1985), xiv.

'fearing that the use of salt would injure their caste'.[9] In this instance, it was Indian merchants and the marketplace that fell under suspicion, but in other cases fears arose that poisonous or ritually polluting substances were being used by the British to kill or incapacitate Indians or, by robbing them of their caste and religion, force their conversion to Christianity. Rumour abounded on the eve of the Indian Mutiny and Rebellion of 1857, at a time of mounting unrest, that food sold in the bazaars had been deliberately tampered with by mixing bone dust with wheat flour in order to break Hindus' caste. Reporting on events in Lucknow in 1857, M. R. Gubbins, Financial Commissioner for the recently annexed province of Oudh (Awadh), remarked that such unsettling rumours were whispered about perpetually and 'the public mind was never allowed to rest'.[10] In July that year, similar reports circulated in the town of Bhopal in central India, alleging that bone dust from cows and (to equal Muslim consternation) from pigs had been mixed with sugar and ghee, and that the English had plotted this desecration.[11] Fear of contaminated foodstuffs added weight to the mounting alarm created by the 'greased cartridges', smeared with pig or beef fat, handed out to sepoys of the Bengal Army in the summer of 1857 and which helped precipitate the mutiny. Poison panics and pollution scares were, Kim Wagner has noted, 'symptomatic of a much wider suspicion of the British'.[12]

Later in the century, there were further eruptions of poison rumours, occasioned this time by suspicions of Western medical and sanitary intervention. The use of potassium permanganate to disinfect wells against cholera, and the aggressive measures taken by the colonial authorities in the 1890s and early 1900s to segregate or inoculate those suspected of bubonic plague, again sent rumour flying, occasioning temporary panics, sporadic rioting and assaults on Europeans.[13] Indeed, in nineteenth-century popular perception in India, the commonest 'folk devil' was often the colonial state and its seemingly sinister agents   doctors, inoculators, ambulance men, Europeans and Eurasians acting in a strange or threatening manner. To judge by

[9] Robert Rankine, *Notes on the Medical Topography of the District of Sarun* (Calcutta: G. H. Huttmann, 1839), p. 20.

[10] Martin Richard Gubbins, *An Account of the Mutinies in Oudh, and of the Siege of the Lucknow Residency* (London: Richard Bentley, 1858), 86.

[11] K. D. Bhargava (ed.), *Descriptive List of Mutiny Papers in the National Archives of India, Bhopal* (2 vols, New Delhi: National Archives of India, 1960), 2: 9, 13.

[12] Kim A. Wagner, *The Great Fear of 1857: Rumours, Conspiracies and the Making of the Indian Uprising* (Oxford: Peter Lang, 2010), 73.

[13] E. H. Hankin, 'Directions for the Use of Permanganate of Potassium in Combating Water-Borne Diseases', *IMG* 31 (1896): 241–47; David Arnold, 'Touching the Body: Perspectives on the Indian Plague, 1896–1900', in Ranajit Guha (ed.), *Subaltern Studies V* (Delhi: Oxford University Press, 1987), 55–90.

popular report, poisoning was what Britons did to Indians, not, as Western Orientalist representation insisted, the other way round.[14] Perhaps, according to popular belief, the aim was to dupe Indians into joining the army or being shipped off overseas as indentured labourers or, more cynically still, to cull an excessively numerous and unwanted 'native' population. One reported explanation of Europeans' malign intent was to capture Indians (the fatter and blacker, the better) in order to extract from them a fluid called *momiai* which could then be used as a salve to save European lives. As in Simla on the eve of the 1857 revolt, a *momiai* scare might provoke panic, triggering the flight of labourers and domestics.[15]

India's poison rumours had much in common with similar episodes in Europe – such as the 'Great Fear' of 1789 and the suspicions surrounding doctors' attempts to contain (or possibly spread) cholera in France in the 1830s.[16] Poison panics have occurred in many different societies, and are not the product of pre-modern societies alone. But, because of their acute sensitivity to anything that might imperil their caste status and religious identity, Indians were particularly susceptible to scares about the purity of the food and drink they consumed and about the bodily violation and loss of ritual status caused by inappropriate touching, drinking and eating. As later discussion will show (Chapter 6), poison, pollution, adulteration and contamination all existed within the same broad spectrum of social concern and collective unease. In British India, such panics were perhaps more likely, too, to be occasioned by a deep suspicion of colonial rule rather than by the class fears that often underpinned such episodes in Europe (though, in West as well as East, xenophobia could help generate poison fears and rumours, like those directed against Jews during the Black Death in fourteenth-century Europe).[17] And, over time, as we will see, poison scares in India segued into public, more especially middle-class, consumer fears about adulterated food and polluted air and water.

---

[14] However, in an age of growing imperial anxiety, it was still possible for medical officers to believe that such rumours were not the result of the untutored imagination of 'ignorant people' but the work of 'interested agitators', who should be tracked down and punished: J. A. Cunningham, 'A Note on the Suppression of Cholera in a Famine Camp', *IMG* 35 (1900): 386.

[15] Gubbins, *Account*, 87; F. S. P. Lely, *Suggestions for the Better Governing of India* (London: Alston Rivers, 1906), 28–29; Evan Maconochie, *Life in the Indian Civil Service* (London: Chapman & Hall, 1926), 83.

[16] R. Baehrel, 'La haine de classe en temps d'épidémie', *Annales: économies, sociétés, civilisations* 7 (1952): 351–60; George Lefebvre, *The Great Fear of 1789: Rural Panic in Revolutionary France* (New York: Pantheon Books, 1973), 62; François Delaporte, *Disease and Civilization: The Cholera in Paris, 1832* (Cambridge, MA: MIT Press, 1986), 48–58, 135–36.

[17] Philip Ziegler, *The Black Death* (Harmondsworth: Penguin, 1982), 98–111.

But, as in 1830s Bihar, there were short-lived panics in which colonialism played no obvious role. To take another instance, there was consternation in Calcutta in August 1910 over an episode of poisoning
(involving half a dozen fatalities) apparently caused by the everyday
practice of chewing betel nuts. As so often the case, the failure to establish
with any certainty the likely perpetrator made poisoning appear all the
more mysterious, sinister and conspiratorial. In the betel nut case, panic
was fuelled not only by the circulation of popular rumour but also, more
in line with Cohen's 'moral panics', by reports in the city's newspapers,
which in turn were accused by the government of 'spreading and keeping
up the scare'. But what was perhaps most revealing about this particular
episode was how quickly the Government of Bengal (already troubled by
widespread resistance to the partition of the province in 1905) moved to
dispel public alarm by setting up a committee of medical experts to
investigate. Within a fortnight it had issued its findings, showing that
there were no grounds for fear and that, if any illness had been caused,
it was attributable to a not uncommon side effect of betel chewing.
Science was mobilized to quell alarm and reassure a frightened public.[18]
By the twentieth century, this use of science as a stabilizing force was
becoming common. In October 1946, more than thirty years after the
betel nut scare in Calcutta, Bombay was shaken by another poison panic.
Rumour circulated that poison had been found in sweets, vegetables, fruit
and other foodstuffs. At a time of mounting communal unrest and political agitation in the city, this was a particularly inflammatory report,
hinting at some Hindu or Muslim conspiracy. But investigation of 200
suspect items by Bombay's Chemical Analyser found no trace of poison.
'It was nothing', he reported, 'but a scare created by some mischievous
persons' who had injected 'some harmless colour into the eatables creating a suspicion of poisoning and making people feel panicky'.[19]

Princely poisoners

Poisoning implicated people, not just substances. During the nineteenth
century, poisoning in India became associated in British minds with
specific social groups and suspect social entities. We have already seen
in Chapter 2 how these might include *vaids* and hakims, practitioners of
Indian medicine who were thought to be reckless in prescribing toxic
substances, and *pansaris*, who sold dangerous drugs in the marketplace.
But alongside these purveyors of 'bad medicine' were other groups

---

[18] Bengal, Municipal (Medical), nos 25–26, September 1910, IOR.
[19] *RCA (Bombay), 1946*, 8; *ToI*, 16 October 1946, 5.

considered to have a close and inappropriate connection with poison practices of one kind or another. Among these were India's princes, thugs and robbers, itinerant fakirs and sadhus, and untouchable Chamars. Each of these suspects groups – significantly ranging from the highest to the lowest in Indian society – attracted the concern, and occasionally the wrathful indignation, of colonial officials, but these diverse social entities also served, by dint of their toxic associations, as means by which to personify evil, to denigrate India, and so endorse the superior moral stature and humanitarian ideals the British claimed for their rule. There was nothing 'civilized' about the way in which poison was seen to operate in India: rather the reverse.[20] And yet no single poison episode or group of poison suspects was sufficient – in the short term – to generate enough of a colonial 'moral panic' to propel the British into adopting more extreme measures for its extirpation. The reliance upon routine science – as demonstrated by the betel nut and foods scares in Calcutta in 1910 and Bombay in 1946 – rather than on panicky responses can be adduced as one reason for this. Other explanations for this relative inertia or contained panic on the part of the state will become more apparent over the course of this and the following chapters. For the present, it can simply be restated that poison was often seen (as with respect to female infanticide or 'criminal abortion') more as a means to an undesirable end than as a cause of mischief in itself, and so the primary direction of colonial policy was to seek to address the underlying source of deviance or criminality rather than to tackle poison, merely its instrument and agency. Poison and poisoning were, in effect, of greater value to the colonial order as a rhetorical device, a means to disapprove and censure, rather than as a primary site of state intervention. Only late in the nineteenth century did that attitude change.

Despite the rapid expansion of British power during the early and middle decades of the nineteenth century, close to a third of the land area of India remained by the 1870s under the notional rule of Indian princes. These were the descendants of once-independent rulers, whose autonomy had been diminished as British supremacy grew and whose activities were kept under the watchful eye of British political agents or 'Residents'.[21] One of the consequences of the 1857–58 rebellion was the decision by the British to halt the annexation of the remaining states and to retain them as a conservative bulwark against future insurrection. Loyal princes were a political and ceremonial asset, but their image

[20] Cf. Burney, *Poison*, 12–16.
[21] Michael H. Fisher, *Indirect Rule in India: Residents and the Residency System, 1764–1858* (Delhi: Oxford University Press, 1991).

remained largely negative. Historical narratives like James Tod's *Annals and Antiquities of Rajasthan* in the 1820s and reports across the century from disapproving Residents alike presented poisoning as an almost routine practice among India's princes, an association that helped sustain the prejudicial idea of the barbaric and anachronistic nature of princely rule.[22] One of the more bizarre examples of this connection between princes and poisons was an attempt in 1875 to assassinate the Maharaja of Bikaner by means of poison concealed in his shoes.[23] That this was not just a colonial perception of princely depravity can be seen from the death in 1883 of Dayananda Saraswati, founder of the Hindu reformist organization the Arya Samaj. He was allegedly poisoned by his cook but at the instigation of a courtesan at the court of the Maharaja of Jodhpur, whom the swami had urged the prince to abjure.[24]

The British accepted that such malign practices resulted in most instances from rivalries, intrigues and the loss of favour within princely families and their court circles and did not necessarily threaten either the Residents themselves or the colonial supremacy that they embodied.[25] But some poison cases were of more pressing concern and suggested how poison might be deployed, not by slaves against their masters, but by Indian princes against their imperial overlords, in what (to British eyes) was a particularly cowardly and treacherous form of anti-colonial resistance. Thus, in 1806, Colonel Colin Macaulay, Resident at the court of Travancore in south-west India, claimed that the state's *dewan* (chief minister) had tried to poison him. The allegation came amidst a flurry of counterclaims that Macaulay had acted high-handedly in trying to enforce Travancore's treaty obligations to pay a hefty annual subsidy to the Company.[26] As so often with allegations of this kind, Macaulay produced no evidence to support his claim but it formed part of an attempt, backed by the Government of Madras, to represent the ruling party in Travancore as inimical to British interests. The episode disclosed imperial as well as personal anxieties about the way poison might be used

---

[22] James Tod, *Annals and Antiquities of Rajasthan* (2nd ed., 3 vols, London: Oxford University Press, 1920); Indrani Sen, '"Cruel, Oriental Despots": 'Representations in Nineteenth-Century British Colonial Fiction, 1858–1900', in Waltraud Ernest and Biswamoy Pati (eds), *India's Princely States: People, Princes and Colonialism* (London: Routledge, 2007), 30–48.

[23] 'Report on the Political Administration of Rajpootana, 1875–76', V/23/40: 215, IOR.

[24] For a version of this story, see https://en.wikipedia.org/wiki/Dayananda_Saraswati.

[25] British anxieties went back to at least 1783, when Tipu Sultan of Mysore was suspected of ordering the poisoning of captured British officers: James Forbes, *Oriental Memoirs: A Narrative of Seventeen Years Residence in India* (2nd ed., 2 vols, London: Richard Bentley, 1834), 2: 459–61.

[26] Robin Jeffrey, *The Decline of Nayar Dominance: Society and Politics in Travancore, 1847–1908* (New York: Holmes & Meier, 1976), 5–6.

as a covert instrument of diplomacy, to ward off overzealous Residents, and, if not to kill them, then to keep them, politically, at arm's length. In fact, Macaulay was soon after removed from his post, perhaps suggesting some doubt about the veracity of his claims or a desire on the part of the Company to repair relations with Travancore and avoid a needless and costly war.[27] In other instances, where the poisoning, or supposed poisoning, came under scrutiny, it appeared that the target had not been the Resident at all but a member of his household, or it was directed against him as an employer, who callously abused or dismissed his servants, rather than in his official capacity.[28]

Speculation that Sir Thomas Theophilus Metcalfe, British Resident at the court of the King of Delhi, may have been poisoned by Zinat Mahal, the wily and ambitious queen, carried rumours and reports of the politically motivated murder of British political agents up to 1853 and the very eve of the Mutiny and Rebellion.[29] But the most thoroughly investigated case of this kind occurred twenty years later and concerned the Gaekwar of Baroda, ruler of a small but strategic state in western India, and his political agent, Colonel Robert Phayre. Suspecting that the current Gaekwar, Malharrao, owed his accession to the throne to the poisoning of his own elder brother, Phayre made little secret of his determination to unseat a ruler whom he saw as politically inept and mentally unsound. One morning in November 1874, as he took his morning drink of sherbet and pomelo juice (a pomelo is a large green citrus fruit, akin to a grapefruit), Phayre noticed that it had a 'most unpleasant metallic taste' and found a suspicious gritty residue at the bottom of his glass. He threw it away, but subsequent analysis of the residue in the tumbler and the Resident's vomit by Dr Seward, the Residency Surgeon, and by Dr Wellington Gray, Bombay's Chemical Analyser, revealed traces of arsenic and 'diamond dust'. The latter appeared particularly incriminating: who but a prince could afford the labour and expense needed to grind up jewels to dust as an aid to murder? But the drink had been delivered to Phayre by his own servants, and, despite detailed investigation, a special commission appointed to look into the Gaekwar's affairs failed to agree on his personal responsibility for the poison attempt.[30] However, despite the tenuous evidence against him, the Gaekwar was removed from his throne on grounds of misgovernment – the most senior prince to be deposed by

---

[27] Political Department dispatch, 29 September 1809, E/4/904, 282: 203–360, IOR.

[28] F/4/1327: 52490, IOR; F/4/1897: 80633, IOR.

[29] William Dalrymple, *The Last Mughal: The Fall of a Dynasty, Delhi, 1857* (London: Bloomsbury, 2006), 114–19.

[30] *Commission of Enquiry into Charges Laid against H. M. Mulharrao, Gaekwar of Baroda*, R/2 538/316, IOR.

the British in the second half of the century. Phayre, too, who had become increasingly paranoid during his time in Baroda and unsettled by bazaar rumours about his imminent death, was transferred to another post.[31]

At one level, the Baroda case reinforced the view of poisoning as an arcane practice, redolent of obscure power struggles in the feudal relics of India's princely states and barely ten years before the founding of the Indian National Congress in 1885 signalled the formal birth of India's independence struggle. Political poisoning was not a stratagem otherwise evident in the history of Indian anti-colonialism, and yet it was not irrelevant either. Taking Travancore, Delhi and Baroda together, one can see how poisoning could be – or be seen to be – a means by which the princes tried to preserve at least a partial autonomy for their beleaguered states. And with hindsight, the Dewan of Travancore, the Queen of Delhi and the Gaekwar of Baroda can be seen in a patriotic light as having struck a blow for Indian freedom.[32] But, more than any other criminal trial in nineteenth-century India (and the investigation into the Gaekwar's conduct was tantamount to a trial), the Baroda case involved extraordinarily detailed examination of the forensic evidence and technical discussion of the nature (and taste) of different poisons, their symptoms and effects and the most effective means of testing for their presence. One of the issues that most exercised Gray as Chemical Analyser was whether the arsenic found in Phayre's glass corresponded to the sample discovered in the possession of one of the Residency servants: examination under a microscope showed the arsenic crystals to be identical. Another issue which puzzled Gray was the presence of diamond dust in Phayre's pomelo drink, which, in the words of the analyst, rendered this case 'unique in the annals of modern poisoning'.[33] Leaving aside the difficulty of grinding obdurate diamonds to dust (an issue that seemed not to trouble Gray), there was the question of what purpose the diamonds were intended to serve. According to A. H. Giles, writing a decade after the Baroda incident, 'diamond powder' was believed by Indians to be 'the most potent of poisons' and as such was used by some would-be suicides, though it could, in fact, 'only injure mechanically' and had no toxic properties.[34] But adding diamond dust to arsenic seemed a needlessly theatrical gesture. Had it really been the

[31] Caroline Keen, *Princely India and the British: Political Development and the Operation of Empire* (London: I. B. Tauris, 2012), 143; I. F. S. Copland, 'The Baroda Crisis of 1873–77: A Study in Governmental Rivalry', *Modern Asian Studies* 2 (1968): 97–123.

[32] For such a view of Zinat Mahal, see 'A Case of Delhi Poisoning?', *Hindu* (Chennai), 5 April 2004.

[33] *RCA (Bombay), 1874–75*, 6–9; *Commission of Enquiry into Charges Laid against H. M. Mulharrao*, 63–68.

[34] A. H. Giles, 'Poisoners and Their Craft', *CR* 81 (1885): 100.

Gaekwar's intention to kill Phayre, or merely to scare him off? Had the zealous, but increasingly distracted, Phayre perhaps staged the poison incident himself, using the diamond dust as a kind of signature by which to incriminate his adversary?

Investigation into the Baroda incident helped elevate forensic chemistry in India to a political, and ultimately a public, prominence that it had never previously commanded, even in O'Shaughnessy's time. The paradox was that Gray's painstaking and sophisticated investigation failed – as many such investigations did in British India – to make much headway against the convoluted politics of palace intrigue and the unreliable testimony of supposed participants in, or witnesses to, the Baroda poison plot. In this princely drama, science made a bold, but not entirely convincing, appearance. But where science wavered, polemic thrived. The Baroda case gave fresh sustenance to a colonial tale of Indian barbarity, murderous intrigues and the vulnerability of honourable British agents to cowardly plots and secret poisonings. As late as 1909, in one of many British retellings of the episode, the Gaekwar was dubbed a 'poison monster' and accused of exhibiting the 'cunning and daring' typical of the 'Eastern poisoner'.[35]

Before we leave the subject of princes as poisoners, one further case deserves consideration, even though it takes us beyond the nineteenth century, if only because it further underscores the fragility of forensic science. In July 1927, Amrit Kaur, wife of the Raja of Kalsia, wrote to the political agent for the Punjab States claiming that her ailing mother, the Maharani of Nabha, was being poisoned. She intimated that Ripudaman Singh – her father, the maharani's estranged husband and the state's deposed ruler – was responsible as he had been trying to regain property assigned to the maharani after they had separated several years earlier. She further claimed that doctors attending the maharani had clear proof that she was being poisoned.[36] The authorities suspected that there was some truth to the daughter's accusations. The maharani had separated from the Maharaja of Nabha because of his violent and abusive conduct: 'a profligate and a blackguard of the worst type', he had made the maharani's life 'hell in Nabha'.[37] The maharaja had further alienated himself from the British as a result of several 'transgressions' and a dispute with the Maharaja of Patiala, a leading loyalist and the premier Sikh prince in Punjab.[38] In 1923, Ripudaman Singh was forced to abdicate

[35] H. L. Adam, *The Indian Criminal* (London: John Milne, 1909), 230, 236.
[36] Amrit Kaur to Col. H. B. St John, 11 July 1927, R/1/1/677, IOR.
[37] S. R. Mayers, Dehra Dun case diary, no. 26, in ibid.
[38] Barbara N. Ramusack, 'Incident at Nabha: Interaction between Indian States and British Indian Politics', *Journal of Asian Studies* 28 (1969): 563–77.

because of misconduct that included kidnapping and alleged poisoning. The maharani moved to Dehra Dun, but had continued to employ some of her husband's former servants. These included a one-time mistress of the ex-maharaja, who was suspected of administering poison to her employer, the maharani, by means of 'medicine' or *prasad* (religious offering).

Also implicated in this tale was a European, Miss Jessop. A former matron in a Calcutta hospital, for twenty years she had been the maharani's companion in India and on her trips abroad, but that trust appeared to be waning. For her forthcoming visit to Europe, the maharani was intending to take her personal physician, Dr Ada Hetherington, and Miss Jessop, resentful at this loss of favour, was suspected of trying to retaliate by poisoning her employer. According to the police, Jessop was a 'drug fiend' (which a naïve or careless typist rendered 'drug friend' in the official report), addicted to quantities of morphine she could not possibly afford on her modest allowance. The implication was that she and/or the maharaja's former mistress were being paid to poison the maharani. When the latter died in August 1927, the case became a murder enquiry, with the maharaja and his supposed accomplices as prime suspects.[39]

Like the Baroda case fifty years earlier, the death of the Maharani of Nabha seemed to reinforce the idea that Indian princes (albeit with the aid of Indian concubines and the occasional European accomplice) were routinely complicit in the use of poison. But, as with the Gaekwar, conclusive evidence of poisoning was difficult to obtain. Despite seemingly authoritative medical evidence from Dr Hetherington and from Dr W. Burridge, toxicologist at King George's Hospital in Lucknow, post-mortem examination of the maharani's body revealed only tiny traces of a 'slow irritant poisoning', perhaps arsenic or mercury. This, however, could be explained by the mercury-based ointment with which she was being treated for an existing illness and the arsenical soap and bandages applied to her ulcerated skin.[40] The Nabha case, no more than that of Baroda, did not induce a sense of panic, except among individuals, like the maharani's daughter, who were personally involved. But nor could any compelling evidence be produced to incriminate the ex-maharaja. Suspicions about princely poisoning remained, but forensic science seemed powerless to unravel the mystery and make a conviction possible.

[39] E. G. Peel, Deputy Inspector-General of Police, 'Report on the Alleged Poisoning of the Senior Maharani of Nabha', 16 and 20 August 1927, R/1/1/677, IOR.
[40] Post-mortem report by Civil Surgeon, Mussoorie, 4 August 1927, in ibid.

Datura thugi

The political ramifications of poison scares extended beyond the princely states and into British India itself. Aside from the immediate danger that the poisoning of its political agents might pose, British officials were apprehensive that criminal poisoning – if allowed to pass unchecked – might threaten the prestige and security of colonial rule. One of the main sources of this concern was the manner in which during the nineteenth century the twin evils of poisoning and *thugi* became intertwined. In the main, *thugi* was seen as a cult of stranglers: having won the confidence of their victims by posing as harmless, companionable travellers, thugs used a scarf or handkerchief to kill them before robbing them and burying their dismembered bodies.[41] There were, however, references in the colonial literature on *thugi* in northern and central India from 1810 onwards that also associated the thugs with poisoning – or, more specifically, with the use of datura, administered through tobacco, betel, food or drink, to stupefy and immobilize victims before they were killed and robbed.[42] In 1820, in one of the first published accounts of the 'murderers called phansigars', Richard Sherwood, a Madras surgeon, identified the use of a *phansi* or noose to kill travellers, describing those who practised this crime as 'villains as subtle, rapacious, and cruel as any who are met with in the records of human depravity'.[43] But in extending his discussion to those he then designated 'Thugs', Sherwood commented that they robbed and murdered travellers 'either by poison, or the application of the cord or knife' – seeming to suggest that poisoning commanded at least equal significance.[44] He also described highway robberies committed by *bairagis*, Hindu religious mendicants, who fell in with a traveller, accompanied him on his journey and then seized the opportunity to mix 'the seeds of the *Datura* or other narcotic plant, with the hooka or food of the traveller, and [to] plunder him when stupefied or killed by the effects of the dose'.[45]

As British investigation into *thugi* expanded and became more systematic, poisoning assumed a less prominent role: thugs were primarily stranglers, and their use of datura, if mentioned at all, was assumed to be a mere preliminary to the gruesome act of strangling their victims, much as the use of opium might aid infanticide or sati.[46] Colonel William

---

[41] K. A. Wagner, 'The Deconstructed Stranglers: A Reassessment of Thuggee', *MAS* 38 (2004): 931–63.

[42] Giles, 'Poisoners', 102.

[43] Richard C. Sherwood, 'Of the Murderers Called Phansigars', *AR* 13 (1820): 250.

[44] Ibid., 287.   [45] Ibid., 291.

[46] For datura's use by thugs, see Kim A. Wagner, *Thuggee: Banditry and the British in Early Nineteenth-Century India* (Basingstoke: Palgrave Macmillan, 2007), 47, 60, 71, 169;

Sleeman, the individual most responsible for identifying and suppressing the thugs and the head of anti-*thugi* operations from the 1820s onwards, made only minor reference to poisoning in his extensive writing on the subject. Poisoning was not referred to in the thugs' confessions he recorded nor did it occupy a special place in his account of the secret language they were said to employ.[47] However, elsewhere, in the narrative of his travels through central India in the 1830s, Sleeman described how a fakir, guardian of a remote Muslim shrine, had, along with his young son, been poisoned by a gang of *dhutoorea*s or 'professional poisoners' of the kind who 'now infest every road throughout India'.[48] The robbers used datura concealed in the food they shared with their victims to incapacitate the old man and rob him of almost his only material possession – a blanket. The fakir recovered after many hours of unconsciousness, but by then his son was already dead, his body devoured by wolves. For Sleeman, the moral of this grim tale of treachery, dispossession and loss was clear. The 'impunity' with which road robbery by means of poisoning was perpetrated, 'and its consequent increase in every part of India', was 'among the greatest evils with which the country is at this time afflicted'. Since Indians appeared unable to defend themselves against such appalling crimes or were powerless to act against the perpetrators, it was incumbent on the colonial regime to act on their behalf.[49]

For the British, *thugi* was 'moral panic' on a scale far larger and infinitely more horrific than that represented by Cohen's 1960s mods and rockers. The acts committed by the thugs were seen as exceptional in their brutality and barbarity; they were taken as evidence of a deep-seated fanaticism, a murderous depravity, a cowardly guile and duplicity, that was inherently Indian. The suppression of *thugi* became, and for decades remained, a cause through which the British could proclaim the rationality, humanity and superior efficacy of their rule. As Superintendent of the Thugi and Dacoity Department, Sleeman was able to persuade the Government of India to adopt exceptional measures for the detection and suppression of *thugi*, measures that lay outside the customary remit of the law. Embodied in the Thugi Act of 1836, these included provision for the use of confessions and approvers to expose and convict others suspected of the crime so that the testimony of one thug could constitute evidence

idem, *Stranglers and Bandits: A Historical Anthology of Thuggee* (New Delhi: Oxford University Press, 2009), 80, 84, 121.

[47] Giles, 'Poisoners', 104; W. H. Sleeman, *Ramaseeana, or a Vocabulary of the Peculiar Language Used by the Thugs* (Calcutta: Military Orphan Press, 1836), 15.

[48] W. H. Sleeman, *Rambles and Recollections of an Indian Official* (2 vols, London: J. Hatchard & Sons, 1844), 1: 109.

[49] Ibid., 114.

for the conviction of others.[50] By the end of 1835, 1,562 individuals had been convicted of thug offences: 382 of these were hanged and 986 transported or imprisoned for life.[51] As late as 1848, Sleeman continued to press for the retention, even extension, of such extreme measures, believing that without them the old 'Thug Associations' would 'assuredly rise up again' and new ones formed. *Thugi* might be in retreat, but vigilance could not be relaxed.[52]

With no fresh convictions after the early 1850s, *thugi* appeared to have been quashed, leaving some to wonder whether the phenomenon had ever really existed at all.[53] According to Sleeman, now the British Resident in Oudh, even the crime of road poisoning, 'which was a few years ago carried on to a great extent', had also now been 'almost altogether suppressed'.[54] But Colonel Charles Hervey, Sleeman's successor as Superintendent of the Thugi and Dacoity Department, began to direct the attention of the government of the North-Western Provinces and the central government in Calcutta to what he saw as the threat posed by robbery and murder by means of poisonous drugs.[55] This was so comprehensively identified with datura that Hervey coined the term 'datura *thugi*' to describe it.[56] In so doing, he combined two emotive words – *thugi* and datura – which individually already expressed outrage and horror and endowed poison robbery with a distinctly Indian persona and an aura of moral repugnance. Here was a 'moral panic' in the making, albeit one largely confined to the colonial administration. In 1860, Hervey appealed to the government to act urgently to suppress this 'heinous crime' and 'growing evil'. When George Couper, Secretary to the NWP government, responded by describing datura *thugi* as 'a crime most dangerous to society and opprobrious to our rule', Hervey promptly adopted that

---

[50] Radhika Singha, *A Despotism of Law: Crime and Justice in Early Colonial India* (Delhi: Oxford University Press, 1998), ch. 5.

[51] [C. E. Trevelyan], 'The Thugs, or Secret Murderers of India', *Edinburgh Review* 64 (1837): 368.

[52] W. H. Sleeman, *Report on Budhuk alias Bagree Dacoits and Other Gang Robbers by Hereditary Profession and on the Measures Adopted by the Government of India for Their Suppression* (Calcutta: Bengal Military Orphan Press, 1849), 2–3.

[53] [Anon.], 'Poisoning in India', *British Medical Journal*, 17 September 1892, 641.

[54] India, Political Proceedings, no. 34, 24 November 1852, E/4/817, IOR.

[55] Hervey claimed that his concern about 'poisoning for the purposes of robbery' went back to 1847: 'Annual Report for 1851 of the Proceedings of the Dacoitee Department', in *Selections from the Records of Government in the Police Branch of the Judicial Department, Vol. 1* (Bombay: Bombay Education Society's Press, 1853), 38, 59.

[56] Hervey claimed that datura *thugi* was also prevalent in Punjab as well as NWP and crossed between British and princely India: Hervey to Secretary, India, Foreign, 17 July 1861, in *Report of Operations in the Thuggee and Dacoity Department during 1859 and 1860: Selections from the Records of the Government of India Foreign Department, Vol. 34* (Calcutta: Bengal Printing Company, 1861), 2–4.

resounding phrase and used it to bolster his demand for special legislation along the lines of the Thugi Act and an India-wide agency for the crime's suppression.[57]

Hervey may have been the inventor of datura *thugi* but he found willing allies for his cause. In a despatch to the Government of India in June 1862, Couper endorsed Hervey's claims about the enormity of the crime and emphasized how little, in practice, it differed from the now-dormant *thugi*. The perpetrators worked in gangs 'almost as thoroughly organized as the gangs of thugs' had been until their suppression. The 'only difference' was that 'drugs, which will certainly deprive any human being of his senses for a time, and may very probably destroy him, are used instead of the handkerchief. There is no difference in the degree of atrocity.'[58] Couper forwarded the findings of the provincial inspector-general of police, who claimed that the true number of such crimes was bound to be greater than the twenty-eight cases of robbery by poison reported to him in recent months. 'The victims are invariably travellers', he explained, 'who are not missed, and of whose disappearance no information is received'. The crime was committed 'in some spot where discovery of its commission is difficult', and the information given to the police related only to cases where the victims survived: 'we do not hear of [the] many cases where death ensues, but where no one is concerned for the murder, which is easily concealed by floating the body in the river.'[59] In lieu of forensic evidence or criminal testimony, there followed a series of brief narratives – potted poison stories – compiled by Hervey and others recounting, in the space of a few lines, how travellers had been waylaid by strangers, who had won their confidence before drugging them, seizing their property, and leaving them helpless, destitute or dead, by the roadside. When such cases did come to light, as they had recently in Benares, they caused 'considerable excitement and horror'.[60] In other words, the Indian public, too, was becoming alarmed by these treacherous acts and violent robberies.

In urging 'stringent measures', the NWP government presented datura *thugi* as a matter of deep abhorrence. 'No crime', it declared, 'can be more hateful'. It had its origins not in 'passion, jealousy, hatred, or revenge'.

[57] G. Couper, Secretary, NWP, to Hervey, 13 September 1860, and Hervey to Couper, 24 September 1860, in Hervey, *Report of Operations in the Thuggee and Dacoity Department during 1859 and 1860*, 48, 65–68.

[58] Couper to Secretary, Foreign Department, India, 11 June 1862, *Selections from Records of the Government of India, Home, Revenue, and Agricultural Department, No. 167: Papers Relating to the Crime of Robbery by Poisoning* (Calcutta: Office of the Superintendent of Government Printing, 1880), 13–14.

[59] Report of the Inspector-General of Police, NWP, 29 May 1862, ibid., 25.

[60] Inspector-General to Secretary, NWP, 19 September 1862, ibid., 35–37.

Rather, it had as its object 'the unlawful acquisition of property by means of which, though always endangering, and not infrequently destroying life, are used with a cold-blooded deliberation and indifference as to consequences, which distinguish its perpetrators as among the very worst and most dangerous of criminals'.[61] But there remained the persistent problem of evidence that haunted so many real or imagined poison episodes in India. Apart from the narratives of travellers who had been waylaid and drugged, there was remarkably little information to support the claims made by Hervey and the NWP government that theft and murder by drugging were widespread and increasing. In the three years 1864–66, across the whole of British India, the Thugi and Dacoity Department could cite only 416 possible episodes of road poisoning, a tiny fraction of all reported robberies and murders.[62] And yet the very paucity of that statistical data made Hervey all the more adamant that this was a crime whose secret nature, like *thugi* before, demanded close investigation and ruthless suppression.

Despite backing from the NWP government, Hervey failed to persuade the Government of India of his claims. There were several reasons for this. One was that, even if such a crime were prevalent in the North-Western Provinces, it did not appear to affect most other parts of the country, or to anything like the same degree. Besides, the Thugi Act of 1836 had been an exceptional measure, necessitated by a particularly horrendous and well-organized crime: the poisoning of travellers, few of whom apparently died, though a cause for concern, was nothing so horrendous, nor so threatening to state authority. It was questioned whether the poisoners Hervey had identified amounted to an organized conspiracy such as thugs were said to have been, or were simply small numbers of unrelated individuals using the same criminal technique. It was doubted, too, whether they ever intended to kill, rather than stupefy, their victims, knowing the power of the drug they used and the greater penalty courts would impose for deliberate murder rather than accidental homicide.[63] And, perhaps most decisively, since the 1830s, a criminal code had been introduced and a reformed police establishment had been instituted across British India. The Government of India was wary of adopting extreme and possibly unwarranted measures that would offend the rule of law when its existing provisions might prove sufficient for the purpose. The central government was anxious to avoid creating any new agency which might prove a 'terror' to the very public it was intended to

---

[61] Government to Inspector-General, NWP, 13 September 1862, ibid., 25–26.

[62] Hervey, *Report on the Crime*, statements for 1864, 1865, 1866.

[63] T. E. B. Brown, *Punjab Poisons* (3rd ed., Lahore: 'Civil and Military Gazette' Press, 1888), 3–5.

protect. Besides, datura grew almost everywhere – in jungles, in gardens and at roadsides. It was utterly impractical to think of outlawing so common a drug, and one with so many medicinal, religious as well as criminal uses.[64] Datura *thugi* was never 'panic' enough to impel the colonial state to see such poisoning as a serious threat to its security.

As late as 1868, Hervey continued to warn that 'the old crime of thuggee by strangling has been superseded by that of poisoning ... an evil accomplished by secret means at once diabolical and cowardly', but even his successor, Major Bradford, was sceptical of such claims.[65] However, though Hervey's pleas went unheeded, the issue (as we will see further in Chapter 4) did not entirely vanish. For decades, references continued to surface in the medical press and in police administration reports, identifying datura as 'the poison of the thugs' and associating it with the drugging and robbing of 'poor wandering pilgrims and beggars'.[66] In discussing datura poisoning in 1911, Bengal's Chemical Examiner, Chunilal Bose, observed: 'In spite of the extinction of the dreadful stranglers, known as Thugs, we have not seen the last of this class of crime which formed a part of their nefarious trade.'[67] But, the authorities' attention also drifted away from 'professional poisoners' and 'dreadful stranglers' to another suspect group – the 'holy poisoners', the fakirs or 'false fakirs', who, on highways or at religious fairs and festivals, used datura to drug and rob pilgrims of their often scanty possessions.[68]

There always was a dissenting voice against the Orientalizing view that crime – not least poison-related crime – in India was exceptional in nature and bore the hallmarks of a perverse and barbaric criminality unmatched by British, or other Western, experience. In 1852, Charles Raikes, an NWP district officer, lampooned the idea that life and property were in any greater danger in India than they were 'at home'. *Thugi*, he wrote,

if we are to believe the frequent correspondence of the 'Times', is more common in the purlieus of Russell Square [in central London] than in the metropolis of Upper India [i.e. Lucknow]. Amongst us in the East, professional villains are to be found occasionally, who, having performed the ascribed ablutions, proceed to drug the traveller for the sake of his money, but it is not in India that wives mix arsenic in their husbands' tea, in order to gain a few pounds by the funeral ... It is

---

[64] Resolution, India, Home, Revenue, and Agriculture Department, 20 February 1880, in *Papers Relating to the Crime*, 6.

[65] Hervey to Secretary, India, Home, 1 August 1868, ibid., 93; Giles, 'Poisoners', 106.

[66] Nil Rattan Banerjee, 'The Symptoms in Datura-Poisoning, with Notes on Thirty-Two Cases', *IMG* 20 (1885): 209.

[67] *ARCED (Bengal), 1911*, 5.

[68] S. T. Hollins, *No Ten Commandments: Life in the Indian Police* (London: Hutchinson, 1954), 112–19; Kim A. Wagner, ' "Treading upon Fires": The "Mutiny"-Motif and Colonial Anxieties in British India', *Past and Present*, no. 218 (2013): 159–97.

in Surrey, not in Agra, that justices of the peace go to bed with a revolving pistol under their pillow. Highway robbery is far more common on the turnpike roads of Old England, than on the Grand Trunk Road of India.[69]

But Raikes was writing before the Mutiny and Rebellion of 1857–58 cast a new, more fretful, shadow over British India and reawakened fears of a murderous and treacherous population, only awaiting the occasion to poison its foreign rulers or reinvent *thugi*.

During the nineteenth century, there were several occasions, as on the eve of the 1857 uprising or during the plague crisis of the 1890s, when poison panics erupted among the Indian population at large, driven by concerns about the poisoning or adulteration of daily items of food and drink. But many of the poison scares of the period were internal to the colonial administration, as British Residents in the princely states fell victim to poison plots or as fears arose of the half-suppressed horrors of *thugi* being resurrected as road robbery by means of datura poisoning. In this almost-spectral form, poisoning represented to the colonial mind the dark side of Indian civilization – the hidden dangers, the murderous duplicity, the indeterminate zone between loyalty and betrayal, subservience and subversion. In the main, though, the Government of India did not believe that these sporadic episodes and periodic alarms were sufficient to warrant the kinds of extreme measures that had formerly been deployed against the thugs or, from the 1870s, against the 'criminal tribes'. It continued to regard poisoning as a crime that could be contained and countered through the existing provisions of the Indian Penal Code or through the professional expertise and forensic skill of the provincial chemical examiners. Poison crime data, like the case studies routinely presented in the examiner's annual reports, were exemplary rather than definitive. They hinted at the nature and extent of a problem of poisoning that was believed to be more extensive than could ever be proved by statistical data or forensic evidence. Anecdote and alarm sometimes took, or threatened to take, the place of more sober analysis, though they could also be met with a deserving scepticism and a quest for more measured responses. The panics occasioned by poison were countered by a strengthening of the colonial information order and an increased reliance upon science – forensic toxicology and medical jurisprudence – to provide, as far as circumstances allowed, a more effective means of detection and control.

[69] Charles Raikes, *Notes on the North-Western Provinces of India* (London: Chapman & Hall, 1852), 187. On arsenic poisoning in Britain, see Burney, *Poison*; Katherine Watson, *Poisoned Lives: English Poisoners and Their Victims* (London: Hambledon Continuum, 2004).

But, given the fallibility of poison science, this was no easy task. In part, this came down to the problem of how to obtain reliable evidence. Despite concerns about princely poisoners like the Gaekwar of Baroda, or the assertions made by Colonel Hervey for widespread datura *thugi*, poisoning remained in many respects elusive, insubstantial, difficult to detect, transient even in its corporeal traces, even when specific cases came to the attention of the police, the magistrates and the chemical examiners. Yet, over the course of the century, there appeared to many British officials and to a wider imperial public to be mounting evidence that poisoning was more widespread than the meagre statistics suggested. In 1868, the deputy coroner for Madras observed that reported cases of murder by poison were 'exceedingly rare', and yet 'in perhaps no other part of the world' did there exist 'greater facilities' for the perpetration of poisoning. Poisons – arsenic, nux vomica, opium – were sold in the bazaars, 'in any quantity, and unrestrained by license'. Other 'deadly poisons', like datura, grew in gardens or flourished by the roadside. Given this easy access, there could be little doubt that the opportunity was taken 'to commit murders which are never recorded as such'.[70]

Poisoning was starting to matter – as much for what was feared as for what was known. In the course of these investigations and deliberations, and through the rise of forensic science and medical jurisprudence, poisoning was beginning to emerge not just as an adjunct to a particular crime (female infanticide, say, or highway robbery), or as testimony to the occasional misuse of a medicinal drug or half-toxic aphrodisiac. Poisoning was beginning to be conceived of as a phenomenon in its own right, a danger, inherent in (and idiomatic of) India, and one that needed, with increasing urgency, to be confronted and resolved.

---

[70] R. S. Mair, *Statistics of Unnatural Deaths in Madras and Other Presidencies and Provinces of India* (Madras: Gantz Brothers, 1868), 12.

4   Toxic evidence

Despite failing in the short term to convince the Government of India, Colonel Hervey's alarm over datura *thugi* contributed to a growing disposition in colonial governance to identify crime with 'habitual' criminals. This reflected a wider trend in Western criminology towards identifying and isolating the 'dangerous classes'. But in India the policy assumed characteristics of its own, being directed against castes and communities whose propensity for crime was seen to be historic and hereditary, tied to their caste identity and customary livelihood. One reason why the Government of India opposed special measures against datura *thugi* was that in the early 1870s, after years of deliberation, legislation was introduced to regulate and police communities officially declared to be 'criminal tribes'.[1] The North-Western Provinces and Punjab, two areas where Hervey had judged datura *thugi* to be rife, were among the provinces where the Criminal Tribes Act of 1871 was most widely and rigorously enforced. Poisoners, many of whom were suspected of belonging to such criminal communities, would, it was hoped, be caught in this capacious net without the need for measures directed against them alone.[2] Poisoning still seemed to be a second-order issue.

And yet Hervey had helped inspire a story of outrage and peril that others took up, embellished and made a matter of public, not merely state, concern. The main vehicle for this renewed emphasis on the dangers of poisoning in India was the growing body of work on medical jurisprudence. Here, too, as already alluded to in relation to princely poisoners and datura *thugi*, questions of poisoning were intimately bound up with issues of evidence. There were two main strands to this. One was the question: to what did poisoning attest? Did poisoning merely

---

[1] Sanjay Nigam, 'Disciplining and Policing the "Criminals by Birth", Part 1: The Making of a Colonial Stereotype – The Criminal Tribes and Castes of North India', *Indian Economic and Social History Review* 27 (1990): 131–64; Mark Brown, *Penal Power and Colonial Rule* (Abingdon: Routledge, 2104).

[2] E. J. Gunthorpe, *Notes on Criminal Tribes Residing in or Frequenting the Bombay Presidency, Berar and the Central Provinces* (Bombay: 'Times of India' Steam Press, 1882), ch. 10.

provide evidence of an individual crime and the specific circumstances of its commission? Or did it testify to a wider culture of criminality and to a peculiarly Indian – and hence racial and communal – weakness and degeneracy? The second issue was more technical and less overtly polemical: how, in the adverse social and environmental circumstances of India, could evidence of poisoning actually be obtained and made answerable to scientific analysis? How, given the material elusiveness of poisoning and the secrecy that enveloped it, could poisoning be made visible, not just to the practitioners of science, but to the critical scrutiny of the state, the law and the public?

The rise of medical jurisprudence

As discussed in Chapter 2, W. B. O'Shaughnessy pioneered medical jurisprudence in India. But his approach to the subject had been through the investigation of the individual cases referred to him and as revealed by techniques of post-mortem examination and forensic chemistry. There was no sense in his work of poisoning being the hallmark of entire social groups or demonstrative of a peculiarly Indian criminality. Besides, after the initial burst of activity under O'Shaughnessy, medical jurisprudence and forensic toxicology had sunk to a low ebb. Although Calcutta Medical College taught medical jurisprudence from its inception in 1835, proposals for an equivalent teaching post at Bombay's Grant Medical College failed repeatedly to win approval in the 1840s and 1850s on the grounds that it was an unnecessary expense.[3] A report in 1852 by a Monsieur Marcadieu in temporary charge of the Calcutta chemistry department showed it to be lacking in basic laboratory equipment and devoid of essential samples of minerals and plant alkaloids – despite the 'vast development' especially of inorganic chemistry in recent years.[4] In 1853, the Governor-General, Lord Dalhousie, informed the Court of Directors in London that the posts of both chemical examiner and professor of chemistry in Calcutta were vacant. Since he knew no one in India qualified to fill them, he asked that a candidate be found in Britain. This was done, but Dalhousie's rationale for having an expert chemist in Calcutta was in relation to the expanding needs of the Geological Survey for expert analysis of coal, iron ore and other mineral samples. Nowhere in his despatch were forensic toxicology and medical jurisprudence even mentioned.[5]

[3] Bombay, Home, to Court of Directors, London, 18 October 1850, F/4/2400: 129419, IOR.
[4] Report of M. Marcadieu, 29 July 1852, F/4/2508: 142613, IOR.
[5] Dalhousie to Court of Directors, London, 3 May 1853, F/4/2526: 145562, IOR.

The revival (and transformation) of medical jurisprudence came from a different direction. In the 1850s and 1860s, there was a marked ethnological turn in colonial epistemology and governance, a growing tendency (even before the outbreak of the rebellion in 1857) to seek a more dependable knowledge of India and Indians through systematic attention to what seemed most conspicuously to define and delineate Indian society – race, religion, caste and tribe. This ethnological imperative expressed a mounting sense of colonial frustration over attempts to reform, improve and modernize India, as well as, after 1857, a pragmatic need to repair and augment a clearly flawed 'information order'. Ethnology was hailed as being 'interesting . . . in a scientific view', but, still more, as 'of vast practical magnitude' to colonial governance in India.[6] And where earlier ethnological studies focused almost exclusively on physical attributes, religion, language and custom, later works gravitated towards criminology, medical jurisprudence and, in consequence, poison.[7]

The leading exponent of the ethnological approach to Indian medical jurisprudence was Norman Chevers of the Bengal Medical Service. He first drew attention to the apparently widespread crime of datura poisoning in an article published in 1854, three years before the mutiny. He elaborated this into a full-length book, published in the same year, which was revised and enlarged several times before his death in 1887. With each successive edition, the scope of Chevers' text grew wider – from its origins in Bengal, it expanded to encompass the North-Western Provinces and finally the whole of India. Like Hervey, Chevers believed that, with the extinction of the old 'profession of Thuggee', the drugging of travellers, with a view to robbery and murder, was becoming 'an established practice among criminals of both sects [i.e. Hindus and Muslims] in every part of the country'.[8] The crime of 'thugi by poison' constituted a great danger to life and property, but, still more significantly for Chevers, it was indicative of the 'criminal characteristics of the people of India' and their 'great moral defects'. Echoing the prejudicial sentiments of writers like T. B. Macaulay and his assertion that Bengalis were 'feeble even to effeminacy',

---

[6] R. G. Latham, *Ethnology of India* (London: John van Voorst, 1859), iii.

[7] Mr Justice Campbell, 'The Ethnology of India', *JASB* 35 (1866): 1–152. On ethnology and colonial governance, see Nicholas B. Dirks, *Castes of Mind: Colonialism and the Making of Modern India* (Princeton: Princeton University Press, 2001), chs 8 and 9; Mark Brown, 'Ethnology and Colonial Administration in Nineteenth-Century British India: The Question of Native Crime and Criminality', *British Journal for the History of Science* 36 (2003): 201–19.

[8] Norman Chevers, 'Report on Medical Jurisprudence in the Bengal Presidency', *IAMS* 2 (1854): 248–49. For Chevers' role in presenting medical jurisprudence as a 'new form of colonial knowledge', see Elizabeth Kolsky, *Colonial Justice in British India: White Violence and the Rule of Law* (Cambridge: Cambridge University Press, 2010), 129–35.

Chevers believed that poisoning was a crime typical of a 'timid people', of a race habituated to duplicity and treachery – the very trademarks of *thugi* – rather than to more overt forms of violence and criminality.[9] Implicit in this was a gendering of both criminal propensity and racial identity – poisoning was what a weak, sneaky, effeminate race did: it was not the characteristic of a bold, courageous, manly people.[10]

Like Hervey before him, Chevers seemed unconstrained by a lack of hard empirical data: indeed, an implicit rationale for the recourse to ethnological generalization and communal stereotyping was precisely the difficulty of obtaining trustworthy, quantifiable evidence. As some contemporaries put it (echoing similar sentiments in the West), what India's medical jurisprudence sought was 'moral evidence', derived from social mores and behavioural characteristics, rather than that found (or not found) through the narrower mechanism of statistics, chemical analysis and post-mortem examination.[11] But there was also a strongly Indian tenor to this moral reasoning. Chevers declared in 1854 that it was essential for India's rulers to have 'an intimate acquaintance with the dispositions, customs, prejudices, and crimes of the people'. The scientific wisdom and forensic experience embodied in Western texts was a useful, but ultimately insufficient, guide: what was required was knowledge of the 'intimate peculiarities of the native character', such as only a 'life-long acquaintance with Hindus and Mussulmauns of all classes' could provide. In this lay a familiar privileging of hard-earned local expertise over remote metropolitan authority. Chevers then proceeded to unleash a list of the crimes he believed characteristic of the Indian people. 'Theft, perjury, personation, torture, child-stealing, the murder of women and of aged men, assassination, arson, the butchery of children for the sake of their ornaments, drugging or poisoning, adultery, rape, unnatural crime, the procuration of abortion' – such were the 'leading villainies of these ingenious, calm-tempered, indolently pertinacious sensualists'.[12]

He later wrote in similar vein:

It is only by thoroughly knowing the people, and by fixing the mind sedulously upon the records of their crimes, that an European can learn how strange a combination of sensuality, jealousy, wild and ineradicable superstition, absolute

[9] N. Chevers, *A Manual of Medical Jurisprudence for India* (3rd ed., Calcutta: Thacker, Spink, 1870), 4–8, 103. On Macaulay and the gendering of race in India, see Mrinalini Sinha, *Colonial Masculinity: The 'Manly Englishman' and the 'Effeminate Bengali' in the Late Nineteenth Century* (Manchester: Manchester University Press, 1995), 15–17.
[10] Chevers' views on the deficiency of Bengali diets are more fully discussed in his *Commentary on the Diseases of India* (London: J. & A. Churchill, 1886), 23–24.
[11] Cf. Theodric Romeyn Beck and John R. Beck, *Elements of Medical Jurisprudence* (5th ed., London: Longman, Rees, Orme, Brown, Green, & Longman, 1836), 664.
[12] Chevers, 'Report', 244–45.

untruthfulness, and ruthless disregard of the value of human life, lie below the placid, civil, timid, forbearing exterior of the native of India.[13]

Poison occupied a cardinal place in Chevers' racial criminology and ethnological jurisprudence precisely because it appeared to be so deeply revealing of Indian – or, in the first instance, Bengali – moral and physical weaknesses. In a lengthy sentence, bursting with prejudicial assumptions, he sought to capture all those evil and exotic factors that, even before the mutiny, made poisoning emblematic of India and the dark, lurking danger to British rule:

The abundance in which a large variety of deadly plants spring up in the hot and moist atmosphere of Bengal, and the unrestricted freedom with which nearly all the most potent kinds of mineral and vegetable poisons can be purchased in every Indian bazaar, added to the familiarity with the action of narcotics which has arisen from their daily habits of opium-eating and hemp-smoking, sufficiently account for the prevalence of the crime of secret poisoning among a timid people who, except when wrought up to a state of frantic excitement, always prefer treachery to violence in the execution of their crimes.[14]

To Chevers, the almost gothic proliferation of poisons in India gave categorical proof of Indian infamy. Arsenic, aconite, nux vomica, opium and *lal chitra* were used for homicide, abortion and suicide; datura and ganja facilitated the indulgent pursuit of intoxication and insensibility; arsenic, copper sulphate, snake venom and *bish* were dispensed as 'medicines in poisonous doses'.[15] The practitioners of India's poison arts were as many and as varied as the drugs they employed – thugs, professional poisoners, abortionists. But, as a physician himself, it was 'ignorant Native doctors', their 'indiscretion and recklessness', their 'poisonous drugs' and 'deadly remedies', that provoked his keenest ire and deepest contempt.[16] He explained how one civil surgeon had seized the stock of a 'native medicinal trader' only to find a ragbag of pseudo-medical junk – alligators' teeth, the dried teats of a jackal, lime, iron rust, and so on – but also, alarmingly, 'several poisons', including arsenic, antimony, datura seeds, ganja leaves and roots of aconite and hellebore.[17] Although Chevers cited British authorities and their texts (hailing Taylor's *Treatise on Poisons* as 'Our highest authority in toxicology'), it was the long litany of Indian criminal cases and the cumulative evidence of Indian poison use and medical malpractice that counted most.[18]

---

[13] N. Chevers, *A Manual of Medical Jurisprudence for Bengal and the North-Western Provinces of India* (2nd ed., Calcutta: Bengal Military Orphan Press, 1856), 8.
[14] Chevers, 'Report', 262.   [15] Ibid., 266.
[16] Chevers, *Manual of Medical Jurisprudence for India*, 154.   [17] Chevers, 'Report', 297.
[18] Ibid., 278.

Chevers was a passionate polemicist, dramatizing the case for an India-specific system of medical jurisprudence in the boldest terms. Such was his influence – and his success in capturing the prevailing mood in colonial science and governance – that a number of later writers followed his lead, further elaborating on the need for an ethnological approach to poison crime and medico-legal evidence. In 1875, Kenneth McLeod of the Bengal Medical Service argued that in India crime followed 'certain grooves determined by custom for nowhere has habit a stronger motive influence than in India. Generation after generation falls into the exact same method of thought, behaviour and life of the preceding one.'[19] This observation, according to McLeod, was more applicable to poisoning than any other form of crime in India, 'and it is fortunate it is so, for the study of the traditions and criminal habits of the people of different provinces and districts, different religions, castes and occupations, as manifested in the annals of crime, will prove a most useful aid to the detection of the case and the particular poison used'.[20] In other words, knowing about the social life and material culture of a particular community – the custom, say, of female infanticide among Rajputs, or for Brahmin widows to be barred from remarriage – could provide the first clue in determining who the likely poison suspect was. Post-mortem findings and the chemical analysis of internal organs and their contents were, McLeod contended, of limited value by comparison: in 1869, he pointed out, only a third of poison cases in Bengal had been determined by means of chemical investigation. Much more was to be gained by considering cases 'in their social and ethnological relations'.[21]

From its inception in 1866, the *Indian Medical Gazette* took up the cause of medical jurisprudence – a medical jurisprudence tailored to India's needs – as one of its principal tasks, publishing numerous articles and editorials on the subject. Many contributors argued that India had physical as well as social peculiarities, effects of climate as well as traits of social conduct, which made crime and its detection substantially different in India from the West.[22] In 1871, McLeod stated: 'All of us who have worked in India know well that the medico-legal aspects of crime differ in many particulars from that of European countries.' He added: 'one does not work long in the country before finding that one has to unlearn much

---

[19] Kenneth McLeod, *Medico-Legal Experience in the Bengal Presidency, Being a Report on the Medico-Legal Returns Received from the Civil Surgeons of Bengal during the Years 1868 and 1869* (Calcutta: Central Press, 1875), 106.

[20] Ibid., 106.   [21] Ibid., 106–07.

[22] On medical jurisprudence, see Michael Clark and Catherine Crawford (eds), *Legal Medicine as History* (Cambridge: Cambridge University Press, 1994); Katherine D. Watson, *Forensic Medicine in Western Society: A History* (Oxford: Routledge, 2011).

and add more to the experience acquired in Europe or from text-books based upon European experience'.[23] An editorial in the *Gazette* in 1889 returned to the 'India is different' argument, citing physical changes that occurred after death, such as the appearance of the skin and the onset of rigor mortis, as substantially different from those observed in 'cold and temperate climates'.[24]

Other medical authors shared this belief in Indian exceptionality, which also, far from being seen as a colonial delusion, received recognition in Britain itself.[25] In the same year as Chevers published his first article on the subject, C. R. Baynes, a judge in south India, wrote his *Hints on Medical Jurisprudence*. This was intended for newly arrived officials who would otherwise have to consult European treatises, 'not one-tenth of which can ... have practical application in this country'. Baynes used Taylor's *Medical Jurisprudence* as a model but adapted it 'as far as possible to the circumstances of this country'.[26] A stream of similar works followed, including in 1889 a textbook by I. B. Lyon, a former Chemical Examiner and Professor of Chemistry and Medical Jurisprudence in Bombay. In this much-cited work, Lyon related the general principles of toxicology outlined by Taylor and Christison to specific Indian examples and relevant sections of the Indian Penal Code. He, too, repeatedly contrasted Indian evidence with British experience – pointing, for instance, to the distinctive use made in India of arsenic, opium, datura and other poisons and in relation to such crimes as homicide, infanticide, abortion, dacoity and cattle poisoning.[27] By the 1880s, Lyon could draw upon a wealth of toxicological information, little of which had been available to Chevers thirty years earlier, including provincial crime statistics, reports from the police and chemical examiners and illustrations from poison trials and such high-profile episodes as the Baroda poisoning

---

[23] Cited in Editorial, 'Our Special Medico-Legal Number', *IMG* 37 (1902): 201.

[24] Editorial, 'Medical Jurisprudence in India', *IMG* 24 (1889): 309.

[25] The 1920 edition of Taylor's text contained a chapter on medical jurisprudence in India (but not on any other part of the Empire) on the grounds that the facts, figures and experiences in India 'do not come within the average experience of practitioners in temperate climates' and presented a number of 'special difficulties'. These included the rapidity with which bodies decomposed, the facilities for concealing or destroying dead bodies (as by cremation), the insufficient particulars of a crime given to the medical officer conducting an autopsy and the 'untrustworthiness of so much native testimony'. W. J. Buchanan, 'A Chapter on Medical Jurisprudence in India', in Fred. J. Smith (ed.), *Taylor's Principles and Practice of Medical Jurisprudence* (7th ed., 2 vols, London J. & A. Churchill, 1920), 2: 886.

[26] C. R. Baynes, *Hints on Medical Jurisprudence, Adapted and Intended for the Use of Those Engaged in Judicial and Magisterial Duties in British India* (Madras: Pharoah, 1854), iii.

[27] I. B. Lyon, *A Text Book of Medical Jurisprudence for India* (Calcutta: Thacker, Spink, 1889), ch. 7.

case.[28] Over time, too, scientific studies seemed to give credence to the idea that India and Indians were, to some degree, different. For instance, by the early twentieth century, there were reported to be a number of small but significant skeletal differences between Europeans and Indians, just as after death cadaveric changes were observed in tropical India that were not entirely identical with such changes in temperate Western conditions. Flies' eggs and maggots – and the most visible signs of corporeal decay – all appeared much more rapidly.[29] Diets were different, and so the evidence that half-digested meals might provide had to take into account what Indians ate and how this might affect the absorption rate of poisons and their post-mortem signs.[30]

Although Chevers' observations and opinions continued to be cited for decades after his death, by the 1870s and 1880s Indian medical jurisprudence had begun to outgrow many of his more sweeping claims and ethnological generalizations.[31] The literature became more nuanced in its handling of social categories, more sophisticated in its discussion of forensic evidence and more willing, too, to accept Indians' own contribution to toxicological enquiry. Among articles published in the *Indian Medical Gazette* in the early 1870s was a series of reports by Robert Harvey of the Bengal Army based on medico-legal returns for the Bengal Presidency. Among other topics, Harvey discussed the number and nature of recent poison cases, underscoring poison's centrality to the subject matter and the methodology of medical jurisprudence at large. For instance, he addressed the familiar issue of datura's criminal use. 'The question', he wrote, 'as to the knowledge by natives of the deadly as distinguished from the stupefying effects of the drug is of great importance in enabling us to judge of the motive and intent of criminals'. He pronounced himself more cautious in this regard than Chevers, believing that poisoners were seldom so reckless as to deliberately seek to murder their victims. He argued, on the basis of legal and forensic evidence, that the number of deaths due to datura poisoning was small, and that poisoners were generally aware of the drug's potentially fatal consequences. Seeing the importance of what Indians themselves believed in such matters, he added that there was a 'very generally entertained opinion, among the uneducated classes – and it is among these classes that most of the

---

[28] On the latter, see ibid., 449. Other examples of this genre include J. D. B. Gribble and Patrick Hehir, *Outlines of Medical Jurisprudence for India* (4th ed., Madras: Higginbotham, 1898); Collis Barry, *Legal Medicine (in India) and Toxicology* (2nd ed., 2 vols, Bombay: Thacker, 1904); Rames Chandra Ray, *Outlines of Medical Jurisprudence and Treatment of Poisons* (2nd ed., Calcutta: The Hare Pharmacy, 1912).

[29] Buchanan, 'Medical Jurisprudence', 888–93.   [30] Ibid., 902.

[31] For a restatement of Chevers' views, see [Anon.], 'Poisoning in India', *British Medical Journal*, 17 September 1892, 641.

poisoners are found – that the plant is an intoxicant and not a poison'. If death resulted from datura, it was likely, therefore, to be accidental, caused by ignorance of the drug's lethal potential, and so deserved a lesser punishment by the courts.[32]

India's medical jurisprudence presents us with a paradox. On the one hand, the work of toxicology and the forensic examination of human remains provided a clear demonstration of the way in which modern bodies, alive or dead, in India as elsewhere, were made answerable to the interrogations of modern science. Bodies were required to submit themselves to science, to give up the secrets, buried within bones and bodily organs, of what was eaten and drunk, how and when individuals had succumbed to a specific kind of poison (and in what quantity) and how such acts of criminality were inscribed deep within the body and not just upon its surface features. And yet, on the other hand, modern science had also to answer to limitations and imperfections that India seemed only to exacerbate.

Imperfect testimony

At the heart of these issues and debates about poisoning, criminality and medical jurisprudence lay the perennial problem of evidence. If we revert for a moment to the Baroda case of 1874, there were two sets of evidential issues involved. The first was whether the arsenic (and diamond dust) found in Colonel Phayre's pomelo juice could be traced to the Gaekwar and so provide evidence of his criminal agency and intent. This is what Wellington Gray, as Bombay's Chemical Analyser, tried, not altogether effectively, to show through his forensic investigation. The second set of issues assumed a wider, far more circumstantial, notion of evidence. Whether or not the arsenic discovered could be traced directly to the Gaekwar, was it likely that an Indian prince would want to poison a British Resident and would he be likely to use arsenic, diamond dust and a bewildering host of servants and intermediaries as the instruments of his crime? Malharrao could not be formally convicted of attempted murder by the first, strictly forensic, route, but the Gaekwar's aberrant conduct – of which the poison plot appeared an indicative part – and even his identity as an Indian prince were in British eyes sufficient to create a presumption of guilt even if they were not sufficient by themselves to have him ejected from his throne. The Baroda case was in some respects exceptional, but it embodied a wider principle of typicality: if it was

---

[32] Robert Harvey, 'Report on the Medico-Legal Returns Received from the Civil Surgeons in the Bengal Presidency during the Years 1870, 1871, and 1872', *IMG* 11 (1876): 115.

characteristic of an Indian prince to use poison (and the historic evidence suggested that it was), so was it characteristic of a Rajput to practise infanticide or a Brahmin widow to abort her unborn child. Cultural values and social circumstance might thus carry an implication of guilt, a presumption of criminal responsibility, more especially in an environment in which other forms of evidence were absent, unreliable or impossibly corrupted.

There were further reasons why poison seemed to exemplify the peculiar difficulties that confronted forensic science and medical jurisprudence in India but also gave it the licence to extend its scientific remit. For instance, there was a widespread belief among colonial officials that Indians' verbal and written testimony could not be trusted. Indians were seen as inveterate liars, given, as Chevers contemptuously put it in 1856, to 'absolute untruthfulness'.[33] In the aftermath of the mutiny, one British artillery officer declared: 'The word of a native, except when it suits his own interests, is never to be believed.'[34] The failure to present a watertight case against the Gaekwar in 1874 partly rested on the inability to elicit consistent and verifiable evidence from Indian witnesses, and this failure was replicated in any number of other Indian poison cases. Shrewd Indian defence lawyers – of whom there was no dearth by the late nineteenth and early twentieth centuries – were not slow to point out inconsistencies and inadequacies in the forensic evidence presented in cases of this kind. They could, besides, avail themselves of Lyon's textbook or one of the many other readily available works of medical jurisprudence to dispute, even ridicule, the findings of the chemical examiner or a police inspector.[35] One resort was, therefore, to look for other means of determining falsehood, guilt or even innocence, and the development of fingerprinting in India well before its adoption in the West was one such technical device used to bypass false witnesses or overturn unreliable testimony.[36] But the forensic skills of the chemical examiner – the evidence provided through post-mortems, the examination of viscera and the use of chemical tests for arsenic and other poisons – offered another (however imperfect) pathway to achieving a similar desideratum.

[33] Chevers, *Manual of Medical Jurisprudence for Bengal and the North-West Provinces* 8; Gribble and Hehir, *Outlines of Medical Jurisprudence*, 15.

[34] George Bourchier, *Eight Months' Campaign against the Bengal Sepoy Army during the Mutiny of 1857* (London: Smith, Elder, 1858), 114.

[35] See the case of Dr Raman, acquitted of causing the death of a patient at the Madras General Hospital: Ranganadha Iyar, *Dr. S. Swaminadhan: A Memoir* (Madras: Hoe, n. d.), 62–63.

[36] Chandak Sengoopta, *Imprint of the Raj: The Colonial Origins of Fingerprinting and Its Voyage to Britain* (London: Macmillan, 2003).

An additional reason for the proliferation of works on medical jurisprudence in late-nineteenth- and early-twentieth-century India was the problematic nature of the evidence itself and the challenging circumstances in which it came to light. Practitioners of forensic science in India had to address and surmount problems of evidence which they saw as peculiar to, or greatly accentuated by, the country and society in which they operated. Forensic science was everywhere beset with difficulties in trying to demonstrate scientifically how and why death and injury had occurred and faced repeated challenges to its methodology and findings.[37] But in India there were held to be still greater problems than in most Western societies. Statements of India's difference morphed into observations about India's defects and deficiencies. In India, where licensed practitioners of Western medicine were rare outside the main towns and cities, many cases of poisoning passed unrecognized or were mistakenly attributed to cholera and other rapid and violent disorders. Responsibility for investigating suspected poisoning cases rested, in the first instance, with the police (though magistrates could also order an inquest to be held). According to Punjab's Chemical Examiner in 1873, the police 'cannot be supposed to be skilled in the observation and estimation of symptoms'. Further, they 'have to get an account of the symptoms after they have occurred, and from ignorant witnesses', and 'to contend with a difficulty in getting them to speak the truth, unknown in Europe'.[38]

Again, post-mortem evidence was said to be 'less definite' in India than in Europe. Until the 1900s, most post-mortem examinations of suspected poison victims were conducted within the districts in which the death occurred by assistant surgeons and civil surgeons, or, as in Calcutta, by surgeons attached to the police department.[39] But since the information sent to the civil surgeon, or forwarded by him to the provincial chemical examiner, was often scanty or even non-existent, the autopsy was frequently carried out 'without any information at all' concerning the circumstances surrounding the death. As a result, the chemical examiner, obliged to work without adequate information about the case, was unsure what he was meant to be looking for. 'This problem', Punjab's Chemical Examiner explained, 'which is seldom met with in a lifetime by an expert in Europe, resolves itself into a search for the poisons commonly used in

---

[37] The problem of medical evidence in India had parallels in Britain: I. A. Burney, 'A Poisoning of No Substance: The Trials of Medico-Legal Proof in Mid-Victorian England', *Journal of British Studies* 38 (1999): 9–92; Katherine D. Watson, 'Medical and Chemical Expertise in English Trials for Criminal Poisoning, 1750–1914', *Medical History* 50 (2006): 373–90.

[38] Cited in T. E. B. Brown, *Punjab Poisons* (3rd ed., Lahore: 'Civil and Military Gazette' Press, 1888), 56–57.

[39] *ARCED (Bengal), 1899,* 5.

the country, unless some suspicious appearances or particles lead to a conjecture in another direction'. The number of poisonous substances (especially vegetable substances) in India capable of causing death was 'practically infinite', and so it was 'impossible with a limited amount of material and time to attempt anything else'.[40]

And then, given the paucity of qualified physicians in the districts and outside the main towns, the nearest surgeon or chemical examiner was often located at a great distance from the scene of the crime. To take an extreme case, Punjab did not have a chemical examiner of its own until a medical college was established at Lahore in 1860: until then, all relevant material had to be sent more than 800 miles to Calcutta for examination.[41] In their annual reports, chemical examiners stressed the importance of stomach contents, viscera and other matter 'liable to decomposition' being thoroughly immersed in preserving spirits, packed in 'perfectly clean glass bottles or glazed jars' and suitably sealed and labelled before they were despatched. 'A minute and detailed account of the suspicions, the symptom, the post-mortem appearances, and the treatment of the case, should in every instance be forwarded.'[42] They seldom were. All too often, the putrid, half-pickled remains reached the examiner in an advanced state of decomposition, 'when the slighter appearances left by disease, injury or poison' were no longer identifiable.[43] In nineteenth-century India, it was often the scientific experts who were mobile and whose travelling gaze was the means by which observations were made and scientific data collected for analysis.[44] In the work of the chemical examiners, it is possible to see a different process at work. The material evidence was itself made to travel and the work of science was made hazardous by the uncertainties of provenance and preservation that movement and distance implied.

There were many other obstacles to contend with. One of the greatest was the widespread custom of cremation among Hindus (at a time when the practice was still rare – and regarded with some abhorrence in the West). This destroyed all, or almost all, traces of poisoning, making it almost impossible, for instance, to determine whether an individual had died (as might be claimed) from cholera or, in fact, from arsenic

---

[40] In Brown, *Punjab Poisons*, 58.
[41] Even in the early 1900s, laboratory facilities in Lahore and Agra were considered very inadequate: Pratik Chakrabarti, *Bacteriology in British India: Laboratory Medicine and the Tropics* (Rochester, NY: University of Rochester Press, 2012), 38–39.
[42] F. N. Macnamara, 'Report of the Chemical Examiner to Government for the Year Ending March 1874', Bengal Medical Proceedings, 172–12/13, 27 May 1874, IOR.
[43] Brown, *Punjab Poisons*, 57.
[44] David Arnold, *The Tropics and the Traveling Gaze: India, Landscape, and Science, 1800–1856* (Seattle: University of Washington Press, 2006).

poisoning.[45] Writing in 1913, Patrick Hehir remarked that the 'great objection' to cremation was that it precluded any possibility of exhumation and so, in effect, favoured the commission of homicide. 'This objection is applicable to the majority of deaths due to foul play, including those from wounds of all kinds and poisoning.' The difficulty, he believed, could only be overcome by 'a minute and detailed *post-mortem* in every case'. But with hundreds of thousands of cremations in India every year, and very few experts to conduct autopsies, that was clearly impossible.[46]

Mortuary records, where they existed at all, were often poorly kept, and, by religious convention, bodies were buried or cremated within hours of death, generally without medical certification as to the cause of death.[47] In the case of a 13-year-old Hindu wife in Calcutta in 1911, it was only when an official at the burning ghat became suspicious, refused to allow a cremation, and had the corpse sent to the police surgeon for examination that evidence of arsenic poisoning – and therefore of murder – was uncovered.[48] There was, besides, intense opposition to post-mortems in India, which were seen as an affront to the dead and a violation of religious rites and duties.[49] Forensic science had to cope, too, with the rapid decomposition of bodily remains in India's heat and humidity, and the speed with which bodies deposited or fallen by the wayside, in jungles or in streams were devoured by tigers, wolves and other wild beasts.[50] According to W. J. Buchanan, the decomposition of a body 'often renders an autopsy not only a difficult, but a very trying, operation for a medical officer'.[51] He explained:

A dead body which has come into the hands of the police has to be carried in the hot weather, covered up with a cloth and wrapped in a piece of bamboo matting, for many miles – thirty, forty, or even sixty – to the nearest medical officer who is alone authorised or competent to make the medico-legal examination.[52]

[45] Gribble and Hehir, *Outlines of Medical Jurisprudence*, 433. By the late 1930s, however, UP's Chemical Examiner, S. N. Chakravarti, established that arsenic *could* be detected in cremated bones and ashes, and so used as evidence of murder: *ARCE (UP), 1939*, 2.

[46] Patrick Hehir, *Hygiene and Disease of India: A Popular Handbook* (3rd ed., Madras: Higginbothams, 1913), 408.

[47] R. S. Mair, *Statistics of Unnatural Deaths in Madras and Other Presidencies and Provinces of India* (Madras: Gantz Brothers, 1868), 3–4.

[48] *ARCED (Bengal), 1911*, 5.

[49] On dissection, see David Arnold, *Colonizing the Body: State Medicine and Epidemic Disease in Nineteenth-Century India* (Berkeley: University of California Press, 1993), 53, 58, 108; Mark Harrison, 'Racial Pathologies: Morbid Anatomy in British India, 1770-1850', in Biswamoy Pati and Mark Harrison (eds), *The Social History of Health and Medicine in Colonial India* (London: Routledge, 2009), 173–94.

[50] D. P. Lambert, *The Medico-Legal Post-Mortem in India* (London: J. & A. Churchill, 1937).

[51] Buchanan, 'Medical Jurisprudence', 886.   [52] Ibid.

But, despite delays and the often advanced state of decomposition, Buchanan insisted that there were few instances in which bodily remains could not be subjected to analysis. He noted that out of 1,300 cases seen by the morgue at Alipore in Calcutta in recent years, only 13 were found to be 'too decomposed' for this to be possible.[53] In 1911, Bengal's Chemical Examiner reported on a case referred to him in which an adult Muslim had reputedly died of cholera. The District Magistrate of Jessore, suspecting foul play, ordered the body to be disinterred and sent to Calcutta for testing. By the time the remains reached the Chemical Examiner, the bones, flesh and internal organs had congealed into an 'indistinguishable' mess. Nevertheless – and here *was* a story of science triumphant – the 'fleshy mass on chemical examination was found to contain arsenic'.[54] There was the added difficulty, alluded to by O'Shaughnessy in his discussion of nux vomica in the 1830s, that some Indians took datura, aconitum, arsenic and other 'poisons' in small but regular doses as tonics and aphrodisiacs. Others were addicted to opium or morphine or, like the Maharani of Nabha, were exposed to arsenic and mercury in the course of their medical treatment. The possibility existed that addiction or medication, and not homicidal intent, lay behind the small but suspicious traces of poison revealed by chemical analysis.

The chemical examiners

The revival of the chemical examiners and their growing scientific and administrative importance can be dated from the early 1870s, when provincial governments began to publish the annual reports of these officers, and some of these, in turn, were reproduced in India's English-language newspapers and so found a public audience. The annual reports provided statistical data on the number and type of cases referred to the examiners, gave brief accounts of the most significant cases investigated and explained some of the methods employed. Thus, in 1874, the Chemical Examiner for Bengal reported that over the previous year his department had investigated 361 medico legal cases, including 186 cases of suspected homicide by poisoning, 99 of cattle poisoning and 45 relating to abortion.[55] But chemical examiners also, to their intense annoyance, received a significant number of body parts where the district medical officer, who performed the initial examination, had failed to identify the

[53] Ibid., 887.   [54] *ARCED (Bengal), 1911*, 5.
[55] Macnamara, 'Report'. Deaths attributed to abortion were frequently referred to the chemical examiners, but the circumstances surrounding abortion made detection extremely difficult: 'Report of the Chemical Examiner to Government for the Year Ending, 15 March 1875', Bengal, Medical, 171–9, 17 April 1875, IOR.

probable cause of death or, unsure of his judgement, had simply passed the material on without offering an opinion of his own, even though there might be clear evidence of strangling, drowning or fatal wounds.[56] In other instances, the samples sent for analysis were so small as to allow only the most obvious poisons to be tested for.[57] In his 1874 report, the Chemical Examiner for Bengal explained the nature of his work by appending a detailed description of the lengthy process by which he sought to isolate and test for poisons. This concluded (after several days of intensive laboratory work): 'If the [remaining] solution contains aconitine it will make an unmistakable impression on the tip of the tongue and portion of the lips to which it may be applied. If the solution contains dhatura, it will, when dropped into a cat's eye, produce prolonged dilation of the pupil.'[58] To judge by this description, there remained something rather impressionistic about forensic science.

Over the years, the volume of work assigned to chemical examiners grew substantially. In the case of Bombay, there was a tenfold increase from only 43 medico-legal cases examined in 1862–63 to 447 in 1877–78. By the 1900s this figure had swelled to over a thousand, and by the 1910s and 1920s reached almost 3,000 cases a year.[59] There were several possible reasons for this increase. In Bombay, provincial legislation introduced in 1866 sought to restrict poison sales: although the act was rather ineffective in securing prosecutions, it may perhaps have made the police and magistrates more alert to possible poisoning cases. Concern over the apparent rise in cattle poisoning may have had a similar effect (see Chapter 6). In addition to their toxicological work, the examiners were required to perform a range of other tasks. In 1873–74, the Chemical Examiner's department in Calcutta investigated 361 medico-legal cases, but it also carried out more than 200 tests on items sent by the arsenal and gun foundry at Fort William as well as drugs, metals and other materials forwarded by government departments. It also analysed samples of water and tested milk for adulteration and impurities.[60]

---

[56] The percentage of cases referred to chemical examiners in which poison was actually detected varied widely. In 1871–72, Bombay's Chemical Analyser investigated 143 cases, of which 100 (70 per cent) provided evidence of poison: *RCA (Bombay), 1871–72*, 3. In 1911, in the same province, in only 36 per cent of cases (113 out of 316) was poison detected, and in 1940, 400 out of 781 (51 per cent): *RCA (Bombay), 1911*, 2; *RCA (Bombay), 1940*, 2. In UP, the detection rate in the early twentieth century fluctuated between 35 and 45 per cent, leading the examiner to complain that 'Very often, bogus and trivial cases are sent up for chemical examination': *ARCE (UP), 1936*, 2.

[57] *ARCE (UP), 1937*, 2   [58] Macnamara, 'Report'.

[59] *RCA (Bombay), 1871–72*, 1; *RCA (Bombay), 1878–79*, 1; *RCA (Bombay), 1911*, 1; *RCA (Bombay), 1921*, 1.

[60] Macnamara, 'Report'.

By the early twentieth century, British India had six chemical examiners. There was one each for Bengal, Bombay, Madras, Punjab, the United Provinces and Central Provinces together, the North-West Frontier Province (following separation from Punjab in 1901) and Burma. Sind had its own analyser, based in Karachi, more than two decades before it was split off from the Bombay Presidency in 1936. It is an indication of the administrative and political importance attached to the post that the chemical examiner reported directly to the provincial government rather than through the head of the medical department, and that, as late as 1927, all but one of the chemical examiners were members of the IMS, still, despite some measure of Indianization, a predominantly European service.[61] In this and other respects, E. H. Hankin in the North-Western Provinces was something of an exception – a bacteriologist who came from outside the prestigious and powerful ranks of the IMS. Trained at Robert Koch's laboratory in Berlin and at the Pasteur Institute in Paris, Hankin remained chemical examiner and government analyst for NWP (later UP) throughout his long Indian career (1892–1922). Hankin conducted research into cholera and plague, devised a simplified test for vegetable poisons and examined ways of preventing the transmission of water-borne diseases. He was also instrumental in the establishment of the Pasteur Institute at Kasauli in 1901.[62] Hankin exemplifies the way in which chemical examiners were often active scientific researchers and not merely state toxicologists.

By 1908, according to one of their number, the duties of the chemical examiner had become 'extremely responsible and arduous, perhaps more so than those of any other appointment open to the [Indian] Medical Service'. This was especially so when the work of being chemical examiner was combined, as in Calcutta, with being professor of chemistry at the local medical college – responsibilities for which he received 'absolutely no remuneration whatsoever'.[63] Such, too, was the growing burden of work and its increasingly technical nature that additional staff had to be appointed to the chemical examiner's department.[64] These assistants were recruited from outside the IMS, thereby opening up new areas of professional employment for Indians with a background more often in chemistry than in medicine. In 1873, Punjab's Chemical Examiner reported that he had appointed a sub-assistant surgeon, Amir Shah, to

[61] J. D. Graham, 'Medical and Research Organisation', in Far Eastern Association of Tropical Medicine (ed.), The Indian Empire (Calcutta: Thacker, 1927), 98.

[62] On Hankin's career, see Chakrabarti, Bacteriology.

[63] J. A. Black to Inspector-General, Civil Hospitals, Bengal, 29 July 1908, Bengal, Municipal (Medical), nos. 9–10, January 1909, IOR.

[64] For example, RCA (Bombay), 1871–72, 1.

help with the growing body of medico-legal cases. The demand for this and other forms of analytical work had made it 'impossible' for one man to oversee all this activity, along with his additional teaching and hospital work.[65] The Indian assistant was often assigned a fairly lowly position, carrying out laboratory tests or office work. In 1873, Amir Shah in Lahore was reported as having been 'as yet employed mainly on analyses of water under superintendence and in microscopic work'.[66] But some chemical examiners paid glowing tribute to the work of their assistants. F. N. Macnamara in Bengal in 1874 thanked Tara Prasanna Rai, saying: 'It would not have been possible for me to carry on the work of the office had I not been supported by his ever ready and intelligent assistance.'[67] By the 1920s, despite periodic retrenchment, most provincial chemical examiners had two or three Indian assistants, many of whom spent their entire careers as assistants to the chemical examiners. But some were very highly regarded and chosen to officiate in the temporary absence of their European superiors.[68]

Indians were also beginning to be appointed as chemical examiners in their own right. In 1915, having served as an assistant or officiating chemical examiner in Bengal for nearly twenty years, Chunilal Bose became Bengal's Chemical Examiner in succession to a series of IMS officers including C. J. H. Warden, L. A. Waddell and C. H. Bedford. Born in 1861 to a high-status Kayastha family in Calcutta, Bose trained at Calcutta Medical College before becoming an assistant surgeon in 1886. The following year, he was appointed Assistant Chemical Examiner for Bengal and Assistant Professor of Chemistry at the Medical College.[69] In a retrospective account of chemical research in Bengal in 1921, the year after his retirement, Bose paid tribute to the work of his European predecessors and referred to his collaboration with several of them including Warren and Bedford. But he was also keen to stress the contribution Indians had made to the field from the time of Kanny Lall Dey in the 1860s onwards.[70] Many Indians of Bose's generation were inspired by the career of the Edinburgh-trained chemist Prafulla Chandra Ray, who discovered mercury nitrite in 1895, established the Bengal Chemical and Pharmaceutical Works in 1899 and published a highly influential *History of Hindu Chemistry*. Ray helped transform chemistry into a

---

[65] *RCE (Punjab), 1873*, 2.   [66] Ibid., 2.   [67] Macnamara, 'Report'.

[68] As in the case of N. J. Vazifar in Bombay: *RCA (Bombay)*, *1925*, 9; *RCA (Bombay)*, *1933*, 10.

[69] J. P. Bose, 'A Brief Sketch of the Life and Career of Chunilal Bose', in J. P. Bose (ed.), *The Scientific and Other Papers of Rai Chunilal Bose Bahadur* (2 vols, Calcutta: Forward Press, 1924), 1: xiii–xv.

[70] Chunilal Bose, 'A Brief Survey of Research-Work in Chemistry in Bengal', in Bose, *Scientific Papers*, 1: 104–12.

prominent field of scientific enquiry for Indians, a form of patriotic as well as public service, and something of this idealistic ethos impressed itself on Ray's contemporary Chunilal Bose.[71] Bose was not alone. After the retirement of Hankin as UP's Chemical Examiner in the early 1920s, he was followed by two Bengalis – Devendranath Chatterji until he retired in 1936, and then S. N. Chakravarti until his death in 1945. In Bombay, when B. Higham of the IMS retired in 1936 after more than twenty-five years as the Chemical Analyser, he was replaced by an Indian, B. Bhujanga Rao, who had previously been Sind's Chemical Analyser. It was a significant mark of bureaucratic and political change that, as in UP, the Bombay post never thereafter reverted to a European or to the IMS.

European or Indian, the chemical examiner became a figure of some public authority. Before 1861, his findings had been treated as inadmissible as evidence in court, but from that time onwards the examiners were increasingly called upon not just to supply written reports but also to serve as expert trial witnesses and give oral testimony in court.[72] By 1911, Chunilal Bose was becoming anxious about the amount of time he was being asked to spend giving evidence in court ('his written reports not being accepted as sufficient') but also uncomfortable about the hostile scrutiny to which his professional expertise was exposed.[73] A practising toxicologist was bound to be cautious about what his evidence had revealed and avoid speculative conclusions about how and why a poison crime had been committed. But this apparent indecision or prevarication did not go down well in court and could be used to undermine his expert testimony. The medical expert was said to make 'a poor impression in court' and was looked upon as 'an unsatisfactory witness'.[74] In response to pressure from Hankin, the UP government stipulated that the chemical examiner should only be required to report on his findings and not to respond to questions of a more general nature, such as the possible cause of death. 'It is the Chemical Examiner's business to ascertain and report facts, and not to draw inferences.' Hankin believed that this was preferable to the procedure in England, where 'one and the same expert testifies to his discovery of a fact and also gives his opinion as to the value of his discovery'.[75]

[71] Pratik Chakrabarti, *Western Science in Modern India: Metropolitan Methods, Colonial Practices* (Ranikhet: Permanent Black, 2004), ch. 8. For Bose's admiration for Ray, see Bose, 'Brief Survey', 108.

[72] *RCA (Bombay), 1871–72*, 2.   [73] *ARCED (Bengal), 1911*, 7.

[74] D. R. Thomas, 'Cases of Poisoning and Suspected Poisoning', *IMG* 76 (1941): 429. For an illustration of the difficulties faced by medical experts in poison cases in the West, see Mark Essig, 'Poison Murder and Expert Testimony: Doubting the Physician in Late Nineteenth-Century America', *Yale Journal of Law and the Humanities* 14 (2002): 177–210.

[75] E. H. Hankin, *The Mental Limitations of the Expert* (2nd ed., Calcutta: Butterworth, 1921), 71–72.

Despite such limitations, the work of the chemical examiners was said by the late nineteenth century to be having a real impact on the detection of crime, and even on its commission. The 'science of medicine' was said to be doing as much as the 'vigilance of the police' in trying to curb the 'evil' of criminal poisoning.[76] A senior Bombay police officer stated in the 1890s: 'Now every native district officer knows precisely what to do. There are fairly competent medical practitioners scattered throughout the country, and it may almost be said that in most cases of suspected poisoning the viscera find their way to the Government Analyst.'[77] He cited the case of a robbery thought to involve poisoning in which the victims' bodies were sent 40 miles from the scene of the crime to the nearest dispensary, where the doctor then removed the internal organs and posted them on, with samples of the travellers' food, to the Chemical Analyser in Bombay. The latter found 'enough arsenic to kill half a regiment'.[78]

Poison and poisoning functioned in a number of different ways in colonial India. At one level, they gave apparent substance to a rhetoric of condemnation, in which the trope of criminal poisoning highlighted what were seen to be the moral weaknesses and material deficiencies of many Indians. A fear of poisoning could also (as will be seen more fully in the next chapter) deeply unsettle the social life and psychology of India's European elite. The claims made for poisoning, especially supposed undetected vegetable poisoning, extended well beyond the available statistical and scientific evidence, and in that very indeterminacy lay part of poison's political potency. And yet, if poison and the fears occasioned by it pointed to apparent deficiencies in the colonial 'information order', the broad response of the regime was to eschew panic and to look instead to science, to routine policing and the law, as the most appropriate and effective means of detection and deterrence.

One aspect of this more measured response was the evolution of a system of medical jurisprudence intended to meet specifically Indian circumstances and characteristics. At its most extreme and polemical (in the hands of Norman Chevers in the 1850s and 1860s), this could amount to a highly political and heavily racialized pursuit, pinning scant facts onto a shaky framework of expansive theories and criminal stereotypes, and in this hostile characterization of race and crime, poisoning served a prominent and exemplary role. But, by the later decades of the nineteenth century, medical jurisprudence in India had mellowed to

---

[76] 'Poisoning in India', 642.
[77] T. C. Arthur, *Reminiscences of an Indian Police Official* (London: Sampson, Low, Marston, 1894), 84.
[78] Ibid., 87.

become a more sophisticated, evidence-based, scientifically savvy endeavour, and one in which Indians as well as Europeans had a formative role. The apparent growth in poison crimes and the rise of medical jurisprudence gave, by the 1870s, fresh purpose and significance to the role of the provincial chemical examiner, whose political status and specialist function had waned since the time of O'Shaughnessy in the 1830s and 1840s. By the close of the century, the post of chemical examiner was again a prominent one, situated at the intersection between state and crime, science and law. And yet, despite the growth in the examiner's duties and resources, the problem of evidence remained, especially the difficulty of providing scientific evidence and giving expert testimony that would withstand scrutiny in the courts. But, as following chapters will show, by the 1890s and 1900s, India's toxic science was beginning to find a new, more public role.

5   Intimate histories

More often than not, poisoning was an intimate act. It happened at home when a family sat down to share a meal, not knowing it contained arsenic or aconite. Family intimates – wives, husbands, lovers, servants, doctors – were among those most frequently implicated in poison crimes or in the accidental administration of toxic substances. Suicide by means of opium or other toxic substances was, with few exceptions, a lonely, desperate act. Even highway robbers depended on a kind of intimacy, as the traveller shared not just the road, but also food and drink, with a seemingly friendly fakir or an innocuous fellow pilgrim. Yet it was also in the nature of poison episodes in India in the nineteenth and early twentieth centuries that they increasingly became affairs of state as well as events freighted with personal significance. The attempt to poison Baroda's Resident, the concern over *datura thugi*, the rise of medical jurisprudence, the accumulating data from the police, magistrates and chemical examiners – all began to engage the anxious attention of the colonial state, even without engendering a sense of panic sufficient to impel special legislation and exceptional executive action. But neither home, nor highway, nor even state adequately defined poison's expanding parameters and growing importance. As the nineteenth century progressed, newspaper reports, court cases and sensationalist literature combined to create a new audience for poisons and poisoning, one that ranged well beyond the purview of science and the political imperatives of the state. Here was an emergent Indian public, for whom even the most intimate acts of poisoning might take on a wider meaning and relevance. It is to this transformation of poison from private fear to public concern that we now turn.

Poison plots

In 1909, two families – the Clarks and the Fullams – in the town of Meerut in the United Provinces (formerly the North-Western Provinces) came to know each other intimately. Henry Clark, aged 41, was a Eurasian assistant surgeon in the Indian Subordinate Medical

Service, a branch of state medical employment confined to 'domiciled' whites and mixed-race Eurasians or Anglo-Indians. His wife, Louisa, six years older, was also Eurasian. She, too, had a medical training, having been a nurse at the Calcutta Medical College hospital before her marriage. By 1909, the Clarks had had four children (only three of whom were still living), but their marriage was a deeply troubled one, erupting into violent and abusive rows over Clark's many flirtations and marital infidelities. The other partners in this Meerut quartet, the Fullams, were domiciled Europeans, born and brought up in India. Edward Fullam was 42 in 1909, and his wife Augusta, whom he had married in 1894, eight years younger. Eddie worked for the Military Accounts Department as a deputy examiner. His father, like Harry Clark, had been a member of the subordinate medical service but Eddie Fullam's position in Military Accounts gave him a salary 50 per cent larger than Clark's, and higher status, while Clark, for want of ability or ambition, struggled to make headway in the medical service.[1]

Fullam's wife, 'Gussie', was the daughter of a Calcutta river pilot. Her family enjoyed a modest respectability in India's capital city, and she was reasonably well educated.[2] But Augusta, 'a society woman fond of social festivities and gaieties', found her marriage loveless and dull, her husband uncaring.[3] 'I am so different', she once confessed. 'I can't live without plenty of love and caresses.'[4] In 1909, the Fullams had two children – Leonard and Kathleen. When a third child, Myrtle (nicknamed 'Carrots' for her red hair), was born in January 1910, Clark was the attending physician: he was probably also the father. Such were the families' increasingly entangled lives that the adults frequented each other's bungalows and participated in the same social events; the Clarks' sons even went to work with Eddie in the Military Accounts Department.

During 1909, Harry and Augusta embarked on a passionate affair, and, despite being transferred to Delhi in November 1910, and then to Agra a few months later, Clark continued to visit Gussie at her home in Meerut. He had previously had casual affairs with European and Eurasian women and sexual relations with those whom Augusta disapprovingly referred to as 'black hens' (Indian prostitutes). She also found repugnant the pornography he wanted to share with her. But, despite this unpromising history,

---

[1] In 1913, the Indian Subordinate Medical Service had a cadre of 450 officers: Clark ranked 13th in the Bengal division: *Quarterly Indian Army List for January 1, 1913*, 550. On the subordinate medical services, see D. G. Crawford, *A History of the Indian Medical Service, 1600–1913* (2 vols, London: W. Thacker, 1914), 2: 100–13.

[2] Molly Whittington-Egan, *Khaki Mischief: The Agra Murder Case* (London: Souvenir Press, 1990), 29.

[3] S. C. Sarkar, *Notable Indian Trials* (3rd ed., Calcutta: M. C. Sarkar & Sons, 1962), 18.

[4] Cecil Walsh, *The Agra Double Murder* (London: Ernest Benn, 1929), 69.

Augusta began to anticipate the blissful life she and Harry would enjoy together once her 'hubby', as she called Fullam in her letters to Clark, had been removed from the scene. But Eddie was still in good health and showed no inclination of wanting to die or to abandon his wife. How much he knew about his wife's infidelity is unclear, though on at least one occasion he found Clark chatting to Augusta, dressed only in her night-clothes, outside her bedroom door at five in the morning.[5] According to her evidence at the subsequent trial, Eddie 'fully suspected and resented her intimacy' with Clark, even threatening to shoot him and then commit suicide, and yet, almost until the hour of his death, Fullam continued to trust Clark as doctor and friend.[6]

The lovers' thoughts turned, inevitably, to poison. It was later claimed that this was simply in order to incapacitate Fullam and hasten his retirement, perhaps to Britain, leaving the lovers unhindered in India. But there is no serious doubt about the lovers' intention to kill him. As Augusta casually remarked in a letter to Clark, 'So the only thing is to poison the soup'.[7] Attempts were also made in April 1911 to poison Louisa by means of arsenic concealed in her food and drink by the family's Indian cook and table servant. When Clark hinted to his Indian Christian friend and confidant Alick Joseph that he was using arsenic to rid himself of his wife, Joseph laughed out loud. He thought of arsenic, as many Indians did, as essentially an aphrodisiac.[8] But, as Cecil Walsh, a former judge of the Allahabad High Court, remarked in his book on the Fullam-Clark case, arsenic had 'always been the prime favourite of the poisoner in India'. It was 'almost tasteless, and is easily available in any bazaar'.[9] There will be more to say about arsenic in the next chapter. But, despite giving Louisa 'enough arsenic to kill ten men', the poison failed to produce the intended effect: since it made her vomit, she ingested very little of it. The disappointed lovers then declared Louisa to be 'poison-proof' and, for the moment, abandoned their efforts to kill her.[10] It seems extraordinary that she remained with Clark despite knowing of his affair with Augusta, and, still more, their attempts to poison her. But Mrs Clark was determined to retain the outward respectability of her marriage rather than seek a very public and shaming divorce, and

---

[5] Ibid., 50–51.   [6] *Times* (London), 17 December 1912, 5.
[7] Walsh, *Agra*, 5, 227–28.   [8] Whittington-Egan, *Khaki Mischief*, 58.
[9] Walsh, *Agra*, 176.
[10] *ToI*, 16 December 1912, 8; Walsh, *Agra*, 41. 'Arsenical poisoning almost always produces vomiting, which in very many cases so far removes poison from the stomach that very slight traces are afterwards found on analysis': Collis Barry, *Legal Medicine (in India) and Toxicology* (2nd ed., 2 vols., Bombay: Thacker, Spink, 1904), 1: 371.

planned to stay with her abusive and negligent husband until their youngest son was old enough for them to live apart.

Harry and Gussie had more success with her 'hubby'. Clark knew little professionally about poisons, though enough apparently to supply the abortion-inducing drugs that enabled Mrs Fullam to terminate at least one of the pregnancies resulting from their affair. But he could easily read about poisons (as he could about abortifacients) in such standard texts as Guy and Ferrier's *Principles of Forensic Medicine* or, better still, Lyon's *Text Book of Medical Jurisprudence for India*, books that were intended to aid the detection of poisoning but could equally serve its commission. Clark even underlined passages in Lyon's textbook and folded down the corners of relevant pages.[11] A 'gambler to the core', Clark was apt to blunder through life, relying more on charm than intellect.[12] As a licensed physician and hospital doctor, Clark had access to poisons from the local chemist's shop, or by ordering through the post from a pharmaceutical firm in Calcutta arsenic, atropine, cocaine and other potentially toxic drugs. Without being fully aware of what was happening, Eddie Fullam began to endure a course of slow poisoning, in which the deteriorating victim appears to be dying from obscure but natural causes. Between April and July 1911, Harry mailed Gussie several packets of poison, which she stirred into her husband's tea or slipped into his tonic Sanatogen. Eddie occasionally complained about a bitter taste and refused to drink any more of the concoction, but otherwise, unlike Louisa, he seems to have had remarkably few suspicions. Harry and Gussie called these poison potions Eddie's 'heatstroke mixture', the intention being to induce symptoms that would be mistaken for apoplexy, a common cause of European incapacity and death in India and one familiar to the families from the fate of their own friends and acquaintances.

The effects of the poison were not all that they intended. Small doses of arsenic were insufficient to kill Fullam outright but they turned his complexion (in Augusta's words) 'a lovely pink' and left him 'full of life and vigour'. Arsenic was widely believed to stimulate sexual desire (which was why it was so widely used in India as an aphrodisiac) with the unwelcome result that Eddie became 'amorous', even 'passionate', towards his wife and would-be killer. 'I have never seen him look better', she reported to Harry with a mixture of disappointment and alarm.[13] The poison plot veered between tragedy and farce. On one occasion, baby Myrtle was nearly killed by grabbing the bottle of poison intended for Eddie and

---

[11] *ToI*, 8 January 1913, 8. William A. Guy and David Ferrier, *Principles of Forensic Medicine* (7th ed., London: Henry Renshaw, 1895); I. B. Lyon, *A Text Book of Medical Jurisprudence for India* (Calcutta: Thacker, Spink, 1889).
[12] Walsh, *Agra*, 251.   [13] Ibid., 80, 180–83.

sucking its cork: 'Carrots' was sick for several days but escaped the fate of many other infants (in India and elsewhere) who died from accidentally ingesting opium, arsenic or other toxic substances left lying around the home.[14] This was far from being 'the fine art of murder' as Thomas De Quincey described it – but then poisoning, an 'abominable innovation', failed anyway to meet his exacting standards.[15]

Eventually the repeated doses of arsenic and a cocktail of other drugs began to produce the desired effect of mimicking heatstroke and apoplexy. His health in evident decline, Eddie was admitted to Meerut's military hospital to recover. When the provincial Medical Board met to review his case in September 1911, it was decided to sanction his early retirement and grant him and his family a free passage 'home to England'.[16] On his discharge from hospital, Eddie's poison nightmare began once more. In a bizarre twist to the tale, Clark sought to acquire a sample of cholera bacilli from Dr Gore, the Assistant Chemical Examiner, in Agra on the grounds that he had devised a cure for cholera and wanted to test it on animals. Gore fobbed Clark off with some 'harmless water microbes', later remarking that he 'evidently knew nothing of bacteriology'.[17] This was not the first or only time that the histories of bacteriology and poisoning intersected. Denied this ruse, Clark found it necessary instead to deliver the toxic coup de grâce to Eddie with a fatal injection of the toxic alkaloid gelsemine. This happened at the bungalow in Agra to which the Fullams had moved only two days earlier. Why the family suddenly relocated to Agra is unclear: perhaps it was because both families were unknown there and the move provided cover for both love affair and murder. Eddie died in his bed on the evening of 10 October, but, as if to underscore the intimacy of this domestic poison drama, the scene was witnessed by Kathleen, the Fullams' 10-year-old daughter. She later testified in court against her mother's lover, just as Louisa's daughter, Maud, gave evidence against Augusta: the Agra murders were a very family affair. Clark then pronounced Fullam dead from heart failure and persuaded Captain Dunne of the Royal Army Medical Corps, who knew nothing of the previous history of the case, to countersign the death certificate.[18]

In a year in which more than sixty cases of suspected poisoning were investigated by the UP police, Eddie Fullam's death attracted little attention. Sudden deaths from heatstroke and disease were not uncommon

[14] William Roberts, 'The Opium Habit in India', appendix to Roberts, *Collection Contributions on Digestion and Diet* (2nd ed., London: Smith, Elder, 1897), 301.
[15] Thomas De Quincey, *On Murder* (Oxford: Oxford University Press, 2006), 26.
[16] Whittington-Egan, *Khaki Mischief*, 99.   [17] *ToI*, 11 January 1913, 9.
[18] Walsh, *Agra*, 210–14.

among Europeans in India, the Fullams were newcomers to Agra, and there was no obvious reason to suspect foul play. His remains were not sent to Hankin, the Chemical Examiner, who in that year alone dealt with 366 cases of suspected poisoning, 61 of them attributed to arsenic.[19] Fullam was buried, rather unceremoniously, the day after his death in Agra's cantonment cemetery, where his grave still lies. After months of plotting, Gussie was relieved at the outcome, anticipating that she would soon be Clark's wife. She wrote to Harry in her usual gushy prose, 'How God has worked out all things so beautifully and brought us two most devoted and loving sweethearts close together, and given us freely to each other!'[20] Caring little by this stage for either his job or his reputation, Clark now spent much of his time at Gussie's bungalow. The lovers' 'criminal intimacy' was fast becoming public knowledge.[21]

In this Shakespearian drama of intrigue, passion and deceit – more *Macbeth* than *Hamlet* or *Othello* – there was still 'poison-proof' Louisa to dispose of. Renewed attempts to poison her failed. Just over a year after Eddie's murder, on the night of 17 November 1912, a band of five *budmashes* (toughs), led by a hospital orderly whom Harry also employed as a servant, staged a fake burglary at the Clarks' home. But the break-in – like so much else in this story – was bungled, and Clark had to return in haste to remove the family's barking dog. Only then could the break-in proceed. Louisa, who had slept through the earlier commotion, was struck on the head with a sword and died the next day. Her daughter Maud, sleeping in an adjacent room, escaped unharmed. It now occurred to the police that Eddie Fullam's death a year earlier had been suspicious, and Clark was arrested on 18 November. The following day a police inspector, investigating Louisa's death, stumbled over a despatch box belonging to Clark under Augusta's bed. It contained nearly 400 passionate and highly incriminating letters from Gussie to Harry, detailing the progress of their affair and the plot to murder their respective spouses. Eddie's body was exhumed on 6 December, and, despite 'more than a year in a cheap coffin in a tropical country', forensic tests carried out by Hankin as the Chemical Examiner revealed only slight traces of arsenic in the femur bones.[22] No evidence of any toxic substance was found in

[19] *Report of the Administration of the Police of the United Provinces, 1911, 11; ARCE (UP), 1911*, 1, 5.
[20] Walsh, *Agra*, 5.   [21] Ibid., 13.
[22] Ibid., 195, 215–16. Details of the exhumation are given in Jaising P. Modi, *A Text-Book of Medical Jurisprudence and Toxicology* (2nd ed., Calcutta: Butterworth, 1922), 66–68. As a lecturer in medical jurisprudence at Agra Medical School, Modi attended the exhumation and took the remains to the Chemical Examiner in person. His textbook also makes a number of references to the Fullam-Clark murders; ibid., 367, 448, 604.

Louisa's body, despite the repeated doses of arsenic to which she had been subjected.

The case against Henry Clark, Augusta Fullam and the *budmash*es for the murder of Edward Fullam and Louisa Clark was heard at committal proceedings at Agra Sessions Court between December 1912 and January 1913. A full trial followed at Allahabad High Court in February–March 1913. Both Clark and Fullam were granted the right to be tried as 'European British subjects', and so, under Indian law, to have a whites-only jury, though it is questionable whether this racial privilege worked in their favour. During the Allahabad trial, Kathleen Fullam gave eyewitness testimony to Clark's involvement in her father's death, including the final lethal injection. In her defence, Augusta claimed to have been captivated by Clark, enthralled by his almost hypnotic hold over her: the murders had been his idea, not hers. Clark then made a statement in which he confessed to being solely responsible for the murders, presumably in order to save Gussie from the gallows. The court also heard a memorandum written by Louisa Clark shortly before her death, listing her many grievances against her husband, his abusive treatment of her and the repeated attempts made to poison her.[23] The exact cause of Eddie Fullam's death could not, however, be definitively established – despite the arsenic traces found in his exhumed body and the forensic evidence presented in court by E. J. O'Meara, Agra's Civil Surgeon. In all, in addition to the report of the Chemical Examiner's Department, ten doctors (from Eurasians in the Indian Subordinate Medical Department to Europeans in the IMS and Royal Army Medical Corps) either gave evidence at the two trials or testified as witnesses to their encounters and conversations with Clark in Meerut and Agra.[24] But, ultimately the medical evidence failed to provide conclusive evidence against the accused even though they were specifically charged with murdering Fullam by administering a poisonous drug. Instead, it was the cache of incriminating letters, together with Clark's confession, that convinced the jury (after only ten minutes' deliberation) of the couple's guilt.[25] A further trial at Allahabad found guilty three of the five assassins charged with Louisa's murder: one of the accused saved his skin by turning approver and giving evidence against the others. Another was acquitted.[26]

After pleas for clemency failed, Clark was hanged on 26 March 1913. But there was evidently anxiety in official circles about executing

[23] A summary of the trials appears in Walsh, *Agra*, ch. 4; Whittington-Egan, *Khaki Mischief*, chs 11–14.
[24] For the medical evidence, see Walsh, *Agra*, ch. 3.   [25] *ToI*, 3 March 1913, 9.
[26] *ToI*, 14 March 1913, 9.

Augusta – a white woman. She was spared the death penalty because she was again pregnant and sentenced instead to penal servitude. She was incarcerated in Allahabad's Naini Jail, a harsh location for a white female who might otherwise have expected to be confined in a European prison in the more temperate climate of an Indian hill station. Her child, born in the jail in July 1913, was taken from her at birth and sent to an orphanage. By what Cecil Walsh considered 'poetic justice', Augusta died in jail on 29 May 1914 from heatstroke, the very cause from which her husband supposedly perished two and a half years earlier.[27]

A crime in context

In some respects, the Agra murder case was exceptional – as much from the attention it received as from the nature of the crime committed. What struck many contemporaries was less a tragic tale of star-crossed lovers, caught up in ungovernable passion, than a rather sordid story of banal and petty lives, in which two unappealing individuals plotted the cruel and calculated murder of their respective spouses. Augusta's letters, described in one press report as a mixture of 'cant, school-girl gush, and penny novelettes', like Clark's promiscuous affairs and serial flirtations, his crass ill-treatment of his wife and the gross misuse of his professional skills, attracted little public sympathy or compassion.[28] At the conclusion of the Allahabad trial, the *Times of India* commented that there had been 'no romance of youth and beauty about the case'. It was 'merely an intrigue between two middle-aged persons ... an intrigue which made up in vice what it lacked in romance'.[29]

Poisoners seldom make public heroes or plausible 'social bandits', though notoriety and the hangman's noose might lend them a perverse fame.[30] The Agra murders quickly became infamous in India and abroad. UP's Inspector-General of Police described the crime as 'almost without parallel in its diabolical atrocity', though he boasted of the Inspector's accidental discovery of Mrs Fullam's letters that it had been 'a most creditable performance of which Scotland Yard itself might be proud'.[31] Double murders, anyway rare in India, became still more sensational when they occurred among the white and mixed-race population. As Walsh observed in his book about the murders, 'except for cases of assassination for political purposes, the murder of either a European or

---

[27] Walsh, *Agra*, 254.   [28] *ToI*, 4 March 1913, 6.   [29] Ibid.

[30] When crowds gathered to witness the execution of the 'Khambekar street poisoner' in Bombay in June 1891, the executioner sold portions of the hangman's rope as a charm against personal misfortune: *ToI*, 8 June 1891, 4.

[31] *Report on the Administration of the Police of the United Provinces, 1912*, 7–8.

an Eurasian is rare, and the murder of an Eurasian woman, in an important and largely populated station like Agra, is almost unheard of'.[32]
According to Walsh, the annals of crime contained 'few stories of passion,
intrigue, and murder, temporarily triumphant, but ending in sudden and
swift retribution, so sordid and, at the same time so remarkable and
engrossing as the Agra Double Murder'.[33] He then allowed his shocked,
but no doubt fascinated, readers to feast on a 250-page account of the
Fullam-Clark murders, quoting at length from Augusta's letters, the trial
proceedings and the medical evidence.

For years thereafter, European and Indian writers eagerly retold this
'Eastern story of passion, murder and intrigue'. For some it was 'one of
the most extraordinary and diabolical murders . . . ever encountered'. 'All
India rang with the horror of it', declared Francis Pearson, author of a
memoir of legal life published more than twenty years after the murders,
adding, with some exaggeration, that the 'whole of India' was 'convulsed'
by the trial proceedings.[34] Extensive accounts of the trials appeared at the
time in English-language newspapers in India and further afield, in
Australia and in Britain, where a correspondent for the London *Times*
declared this to have been 'one of the most remarkable criminal trials in
history'.[35] When a European dentist was shot dead in Travancore in
December 1914, and his wife and her lover were suspected of the crime,
it was at once suggested that the motive was the same as in the 'notorious
Clarke-Fulham [*sic*] murder case'.[36] The story of the Agra murders has
been retold many times since, whether as 'true crime' or as fiction. It was,
for instance, reworked decades later as a short story – entitled 'He Said It
with Arsenic' – by the Anglo-Indian writer Ruskin Bond.[37] Poison tales
always lent themselves to misrepresentation and mythologizing. In his
memoirs, Pearson confused Augusta Fullam with another European
woman imprisoned at about the same time and mistakenly claimed that
Gussie had been released from prison after serving only eighteen months,
married an official at the Rumanian consulate in India and was now (in
the 1930s) living in Bucharest.[38]

But the Agra double murder was not quite as rare and unrepresentative
as it might at first appear or as many contemporaries liked to imagine. It
happened at a time when colonial India was troubled by a series of poison
scares, some of which contributed to the introduction of the Indian

---

[32] *ToI*, 4 March 1913, 6; Walsh, *Agra*, 164.   [33] Ibid., 13.
[34] Francis Pearson, *Memories of a K.C.'s Clerk* (London: Sampson, Low, Marston, n.d. [c.
1935]), 263, 269; cf. Sarkar, *Notable Indian Trials*, 18–26.
[35] *Times*, 11 March 1913, 7; *ToI*, 4 March 1913, 6.   [36] *Madras Mail*, 5 January 1914, 6.
[37] [Ruskin Bond], *The Best of Ruskin Bond* (New Delhi: Penguin, 1994), 124–28.
[38] Pearson, *Memories*, 273.

Poisons Act in 1904 (for which see Chapter 6). The Fullam-Clark murders came too late to impact on that legislation, but an earlier, less publicized case did. In 1895, Ellen Wagner and her accomplice James Cray were suspected of poisoning her husband, a pipe layer, with arsenic. This was one consideration, among many, in finally forcing the Government of India to act against unregulated poison sales in India.[39] In March 1912, again at Allahabad High Court, Eva Stephens was charged with the murder by prussic acid of a psychic, Garnett Orme, against whom she may have had a personal grudge. But, unlike Augusta Fullam, she was acquitted, leaving Orme's death an unresolved mystery.[40] When it came, ten years on, to review the working of the Poisons Act in 1914, the UP government referred to two 'notorious cases' of poisoning in the province in recent years – one the Orme murder, the other the Fullam-Clark case.[41]

In fact, poison cases involving Europeans, whether as perpetrators or victims, were not all that uncommon in British India. Indeed, such recurrent episodes touched (as we will see shortly) on many of the underlying fears and phobias of the colonial regime. Likewise, the failure of forensic science conclusively to establish the use of arsenic to poison Eddie Fullam and Louisa Clark, and, by contrast, the greater reliance placed upon Augusta's letters and Harry's confession, again highlighted the persistent difficulty, alluded to in the previous chapter, of providing convincing evidence for homicidal poisoning. This uncertainty bred a degree of scepticism among magistrates as to the worth of the testimony presented by forensic experts and chemical analysts. In the same year as Eddie's death, a district magistrate in UP warned against 'too slavish an acceptance of medical evidence' in murder cases.[42] The failure to identify Eddie's death as suspicious until after Louisa's killing a year later suggests the frequency with which (even among Europeans) possible cases of poisoning were ignored or too readily ascribed to the effects of climate and disease. Although the Agra murders happened among Europeans and Eurasians, the involvement of Indian servants in delivering cups of tea and tumblers of tonic, knowing them to be tainted with arsenic, is suggestive of the way in which poisoning practices not uncommonly crossed the racial divide: poisons could as well exist as everyday substances in European, as in Indian, homes.[43]

---

[39] India, Home (Judicial), nos 80–81, August 1895, IOR.   [40] *Times*, 19 March 1912, 8.

[41] S. P. O'Donnell, Secretary, Judicial (Criminal), UP, to Secretary, Home, India, 1 April 1914, India, Home (Judicial), no. 265, October 1914, IOR.

[42] *Report on the Administration of the Police of the United Provinces, 1911*, 11.

[43] For instance, in 1891, a European nurse in Poona was poisoned (with arsenic) and robbed by her Indian servant: *RCA (Bombay), 1891*, 5.

Sensational poison cases, like the Agra murders, became a quest after meaning. They gained emotional force and social traction from the way in which they used the terrain of the private sphere and the domestic space of the home to purvey morally edifying and politically empowering stories about public morality and personal vice.[44] The rise of the daily press, in English and in India's regional languages, the publication of popular accounts of murders and trials and the many works of fiction and drama that traded on them – all contributed to the proliferation of poison tales. In India as elsewhere, these fed a growing consumer appetite for such shocking but, in Walsh's words, 'engrossing' stories. Under colonialism, India's emergent public sphere was no more immune than its Western counterparts to the voyeuristic spectacle of troubled lives, illicit affairs, poisonous intimacies, murderous infidelities and the moral dilemmas they so vividly presented to the public gaze.[45] Indeed, the new Indian public and its many vehicles – press, theatre, street songs, popular prints – thrived on such cases and feasted on their notoriety. A character in one of Rabindranath Tagore's short stories of the period 'imagined that most modern-day Bengalis were much like himself – that is, they ate, slept, and had little faith in anything except scandal'.[46]

The sensation created by the Agra murders in white society in India was matched by a number of Indian murder trials in the late nineteenth and early twentieth centuries. Among the most dramatic and best known of these was the Tarakeshwar murder case. In 1873, Elokeshi, a young Brahmin wife, had her throat slashed with a fish knife by her outraged husband, Nabinchandra Banerji. He had discovered that she had been seduced by the *mahant* (chief priest and proprietor) of a temple near her parents' home at Tarakeshwar, north of Calcutta. Poison was not directly involved in this case, but the Tarakeshwar temple had become renowned for miracle cures, including the treatment of female infertility, and the 16-year-old Elokeshi had first gone there seeking 'medicine' to help her conceive. In a society that valued procreation so highly – and the birth of male heirs in particular – and where infertile wives were humiliated, shunned and rejected, there was great demand for fertility drugs. What Elokeshi received was one of those 'aphrodisiacs' that Western

[44] Karen Chase and Michael Levenson, *The Spectacle of Intimacy: A Public Life for the Victorian Family* (Princeton: Princeton University Press, 2000), 6–7.

[45] On the nature and emergence of India's 'public sphere', see Sandria B. Freitag (ed.), 'Aspects of the Public in Colonial India', special issue, *South Asia* 14 (1991); Partha Chatterjee, *The Nation and Its Fragments: Colonial and Postcolonial Histories* (Princeton: Princeton University Press, 1993); Sanjay Joshi, *Fractured Modernity: Making of a Middle-Class in Colonial North India* (Delhi: Oxford University Press, 2001).

[46] 'The Austere Wife' (1917), *Selected Short Stories of Rabindranath Tagore* (London: Macmillan, 1991), 250.

medicine derided as quackery and condemned for containing arsenic, aconite and other toxic ingredients. In one version of the story, Elokeshi was drugged (perhaps with datura, the 'date rape' drug of the time) before being sexually assaulted by the *mahant*. Overcome by remorse, Nabinchandra confessed to his crime and was sentenced to life imprisonment for murder, but, following a torrent of public petitions in support of a husband who had been 'wronged', he was pardoned and released. The *mahant* was sentenced to three years in jail and fined Rs 3,000. The affair of Elokeshi, her seducer and her husband excited enormous public debate in Bengal – about sexual morality, the conduct of wives and the rights and duties of husbands.[47]

Closer in time to the Fullam-Clark murders was the Bhawal Sannyasi case, another episode which could not fail, according to one Indian commentator, to 'grip the reader's mind'.[48] This involved Ramendra Narayan Roy, the young heir to a zamindari estate in east Bengal. He fell ill while staying in Darjeeling in 1909, possibly, so it was later claimed, as a result of being poisoned with arsenic, given as a cure for his colic. In the belief that he was already dead, Roy's inert body was taken to the cremation ground – but, before it could be burned, he was (supposedly) found still alive and rescued by a passing Hindu holy man, who then inducted him into the life of a wandering mendicant. Years later, the putative zamindar returned to claim his stolen inheritance. The mysterious disappearance and enigmatic return of the Bhawal claimant gave rise to a series of court cases in which the medical and forensic evidence again stood centre stage – indeed, the case began with a European physician, Dr Calvert, whose ministration of a medicine, perhaps legitimately containing arsenic, first led to accusations that he had been part of a plot to murder and disinherit the young heir. Eventually, the claimant was exposed as an imposter, but the 'story of poisoning' – the nature of his illness, the size of the arsenic dose he had received, the possible effects of the drug in changing his voice and physical appearance – became central to the long-running litigation and the enormous publicity and partisan passions it aroused.[49] Here, too, fact and fiction, medicine, mystery and poison, were enthrallingly – and very publicly – entwined.

[47] Tanika Sarkar, *Hindu Wife, Hindu Nation: Community, Religion and Cultural Nationalism* (London: Hurst, 2001), ch. 2.

[48] S. C. Das Gupta (ed.), *The Bhowal Case: High Court Judgements* (Calcutta: S. C. Sarkar & Sons, 1941), ix.

[49] Ibid., 69, 85–88, 102–07. See Partha Chatterjee, *A Princely Imposter? The Strange and Universal History of the Kumar of Bhawal* (Princeton: Princeton University Press, 2002).

Colonial intimacies

Gross acts of poisoning, like the Agra murders, reveal much about the social structure and political construction of British India. They offer insight into the function and effect of grand narratives of race, gender and colonial legitimacy. But such episodes also serve, narratively and analytically, to direct attention to the interconnected intimacies of private lives and secret poisons.[50] Something of this connectivity was suggested in Chapter 1 in discussing the 'social life' of poisons in pre-colonial and early British India, but the linkages between private and public, between science and sensationalism, had given poisons and poisoning a new vigour and trenchancy by the early twentieth century and in relation to both British and Indian lives.

In histories of British India – as in kindred forms of colonialism around the globe – intimacy has conventionally been understood in terms of sexual intimacy. It has been identified in particular with sexual transgression across the formal lines of racial division that typified colonial societies and which constituted primary markers of what Partha Chatterjee has termed 'the rule of colonial difference'.[51] The apparent retreat from sexual liaisons between Europeans and Indians by the 1830s has even been seen as presaging the 'decline of intimacy' itself, as if no other kind of intimacy could exist other than that of the bedroom or brothel.[52] But there is no compelling reason why intimacy should be confined to sexual relationships or to the lives of women rather than men. In her seminal work on the Dutch East Indies, Ann Stoler has used intimacy – the 'imperial politics of intimacies' – to indicate the problematic spatial, social as well as sexual dynamics of the relationship between the colonial Dutch and their East Indian (now Indonesian) subjects. As she puts it, 'intimate matters and narratives about them' had a crucial role in 'defining the racial coordinates and social discriminations of empire'. Used in this way, intimacy retains a transgressive quality. It violates, to a degree subverts, the established norms and conventions of a colonial society in which whites and non-whites are – in theory – kept strictly apart. But Stoler's analytical use of intimacy, while retaining some of its transgressive worth,

---

[50] For poison as a crime of intimacy and of 'intimate violence', see Martin Wiener, 'Alice Arden to Bill Sykes: Changing Nightmares of Intimate Violence in England, 1558–1869', *Journal of British Studies* 40 (2001), 184–212; Ian Burney, *Poison, Detection, and the Victorian Imagination* (Manchester: Manchester University Press, 2006), 21–25.

[51] Chatterjee, *Nation*, 10.

[52] Sudipta Sen, 'Colonial Aversions and Domestic Desires: Blood, Race, Sex, and the Decline of Intimacy in Early British India', in Sanjay Srivastava (ed.), *Sexual Sites, Seminal Attitudes: Sexualities, Masculinities and Culture in South Asia* (New Delhi: Sage, 2004), 49–82.

also encompasses the close relationship between Europeans and their 'native' servants, especially the household servants with whom they shared so much of their domestic space and no small part of their private lives. As Stoler puts it – she might almost have had poisoning in mind – such intimacies inhabited the interrelated domain of 'kitchens, bedrooms, and nurseries'.[53]

In a manner similar to Stoler's, in his 'intimate portrait' of late-eighteenth- and early-nineteenth-century Calcutta, Peter Robb has described Richard Blechynden's complex relationships with his European friends and colleagues but also the 'household intimacies' by which he was attached to his children, his mistresses and his household servants.[54] Indian servants might be beaten, abused and dismissed, but, with respect to far more than sex, their lives and their modes of work and leisure remained entangled with those of their European neighbours and employers in ways that blur customary notions of racial hierarchy and social segregation, and defy any neat bifurcation between colonial ruler and 'native' subject.[55] Within the confines of a colonial bungalow, with its open doors and interconnecting verandas, there was little, even among affluent, high-status Europeans, to preserve their privacy and obstruct trans-racial intimacy and interdependence.[56] Here was a porous, permeable space, one in which, as in the Fullam and Clark households in Meerut and Agra, poison might pass from white master or mistress to Indian servants and back again.[57]

Many Indian poison tales are thus remarkably revealing of what Sara Suleri refers to as the 'necessary intimacies' between ruler and ruled.[58] Such stories might have no obvious moral tale to tell, but that did not

---

[53] Ann Laura Stoler, 'Tense and Tender Ties: The Politics of Comparison in North American History and (Post) Colonial Studies', *Journal of American History* 88 (2001): 829–65; idem, *Carnal Knowledge and Imperial Power: Race and the Intimate in Colonial Rule* (Berkeley: University of California Press, 2002).

[54] Peter Robb, *Sentiment and Self: Richard Blechynden's Calcutta Diaries, 1791–1822* (New Delhi: Oxford University Press, 2011), xi, 58–64; idem, *Sex and Sensibility: Richard Blechynden's Calcutta Diaries, 1791–1822* (New Delhi: Oxford University Press, 2011), 14.

[55] Racial violence was often also intimate violence: Elizabeth Kolsky, *Colonial Justice in British India: White Violence and the Rule of Law* (Cambridge: Cambridge University Press, 2010), ch. 1; Durba Ghosh, *Sex and the Family in Colonial India: The Making of Empire* (Cambridge: Cambridge University Press, 2006), 179–91.

[56] On the interior space of the European bungalow, see Elizabeth M. Collingham, *Imperial Bodies: The Physical Experience of the Raj, c. 1800–1947* (Cambridge: Polity Press, 2001).

[57] The idea of Indian servants transmitting invisible poisons to their European employees was reinforced by bacteriological studies from the 1880s onwards showing how unseen diseases might be transmitted in exactly the same way: Pratik Chakrabarti, *Bacteriology in British India: Laboratory Medicine and the Tropics* (Rochester: University of Rochester Press, 2012), 64, 189.

[58] Sara Suleri, *The Rhetoric of English India* (Chicago: University of Chicago Press, 1992), 3.

prevent them from being read and interpreted as moral vignettes or political cautionary tales. At one level, poisoning seemed to be little more than one of the multitudinous hazards of Indian life. It was, for instance, common for manuals of European family health in India to refer, almost casually, to the dangers of accidental poisoning in the home. As noted in Chapter 2, Indian ayahs might use small pellets of opium to keep European children quiet and docile, always raising the possibility of their young charges accidentally receiving a larger, more toxic, dose. But the recommended solution was to hire a new ayah rather than seek recourse to criminal law. Edward Birch's book *Management and Medical Treatment of Children in India*, first published in 1879 and in its fourth edition by 1902, devoted several pages to snakebites, scorpion stings and the treatment of children who had been overly exposed to opium or arsenic in household remedies and patent medicines like Dover's Powder and Fowler's Solution. To judge by such accounts, arsenic might lurk almost anywhere in the European home – in flypapers and rat poison, in medicinal soap and decorator's paint – just as aconite might infiltrate the kitchen, its whitish-grey root mistaken for horseradish.[59] In this respect, perhaps, the European home in India represented nothing more sinister than an exaggerated version of the accidental toxic dangers that haunted middle-class households in Victorian Britain.[60]

And yet, from a European perspective, India was not quite so normal. Poison was seen to pose greater dangers to Europeans in India than 'at home', and served as a signifier of the ways in which Indian criminality, malevolence or simple negligence might impinge directly – even catastrophically – on their everyday lives. In one of the large number of works on crime in India produced for a British readership within a decade or two of the Agra murders, H. Hervey identified poisoning as a 'very common crime' in India, one to which Indians of all classes resorted whenever 'revenge has to be satiated, a score to be paid off, or anyone obnoxious to be got out of the way'.[61] One of the poison cases to which Hervey gave close consideration had certain parallels with the Fullam-Clark murders – in the central role played by Indian servants in a European household and the scenes of domestic intimacy and racial interdependency it summoned

---

[59] Edward A. Birch, *The Management and Medical Treatment of Children in India* (4th ed., Calcutta: Thacker, Spink, 1902), 441–52.

[60] J. W. P. Bartrip, 'How Green Was My Valence? Environmental Arsenic Poisoning and the Victorian Domestic Ideal', *English Historical Review* 109 (1994): 891–913; James C. Whorton, *The Arsenic Century: How Victorian Britain Was Poisoned at Home, Work, and Play* (Oxford: Oxford University Press, 2010).

[61] H. Hervey, *Cameos of Indian Crime* (London: Stanley Paul, n.d. [c. 1912]), 294.

up.[62] At a small civil station in northern India, an ill-tempered European, identified only as Mr F, had a reputation for mistreating his Indian servants. He had more than once been summonsed for assault, but, secure in the privileges of his race, had been punished with only modest fines. On one occasion, however, he dashed a jug of boiling milk in the face of his *khansamah* (butler or table servant), complaining that it had been diluted with water. The servant, a Punjabi Muslim, was severely scalded in this attack and sought revenge by putting powdered glass in his master's favourite dessert, a guava meringue. However, the servant confided his intention to his lover, a Eurasian maidservant, who in turn warned Mrs F to be wary of any food served by the *khansamah*. It was not long before Mrs F spotted something glistening in her husband's dessert. The police were called and evidence was quickly found (as in a number of similar cases) of glass having been ground up on a grindstone in the kitchen. The *khansamah* was sentenced and convicted for intending to cause harm to, and possibly kill, his employer, but the unfortunate maid, who might have been thanked for her timely intervention, was also dismissed.[63] Poisoning, as Megan Vaughan has observed, was often a 'betrayal of intimacy'.[64] In this case, poisoning involved a betrayal of the white master's reliance on his Indian servant and the food he was trusted to deliver to the family dinner table.[65] But the incident also contained other acts of betrayal – the betrayal of the *khansamah*'s trust in the maid and of the maid's confidential revelation to her mistress.

This was just one of many instances in which Europeans suspected their Indian servants of trying to poison them, or of being agents in poison plots in which they became victims. Perhaps Europeans simply transferred to their servants the unease they felt about India at large – alien and therefore poisonous – or sought to blame them for their recurrent but unexplained bouts of illness. But other cases clearly involved more complicated domestic intrigues and rivalries. The Fullam-Clark trials had barely concluded before a second poisoning case opened at Agra in March 1913, this time at the cantonment magistrate's court. Captain Anderson of the Seaforth Highlanders and his wife reported having suffered repeated bouts of giddiness, stomach pain and vomiting. One of their servants, the sweeper's wife Rumpoo, also fell ill after eating

---

[62] Such cases were not uncommon around the time of the Agra murders: *ARCE (UP)*, *1913*, 2.

[63] Hervey, *Cameos*, 300–05.

[64] Megan Vaughan, *Creating the Creole Island: Slavery in Eighteenth-Century Mauritius* (Durham, NC: Duke University Press, 2005), 99.

[65] On the kitchen as a problematic site of European scrutiny and control, see Mary Procida, 'Feeding the Imperial Appetite: Imperial Knowledge and Anglo-Indian Discourse', *Journal of Women's History* 15 (2003): 123–49.

leftovers from the Andersons' plates – itself an indication of the intimacies and hierarchies involved in food preparation and consumption in a colonial household. Mrs Anderson had previously complained of finding mud in her bathwater, and one day her soup was so full of chilli she could barely touch it. The breakfast porridge had also been tampered with. One of the accused, Madar Bux, was a *khansamah* whom Captain Anderson admitted to striking on several occasions for being lazy and impertinent. He had refused to reinstate Madar Bux's father-in-law, Khoda Bux, the family water carrier, and had a grievance against yet another relative, Nur Bux, who had also once served with the Andersons. UP's Chemical Examiner found traces of arsenic in the vomit of all three victims – Captain Anderson, Mrs Anderson and Rumpoo. The main reason for the poison (a 'medicine' supplied by yet another aggrieved ex-servant) appears to have been a plot to incriminate the family cook, so as to have him dismissed and thereby wreak revenge on the Andersons for their ill-treatment of the Buxs. The case against Madar Bux failed for want of evidence, but Nur Bux was found guilty and sentenced to seven years' imprisonment.[66]

It is possible to move from the Agra murders, and other cases principally involving Europeans, to tales of poisoning which, by means of the chemical examiners' reports, enable us to eavesdrop on Indian intimacies. So we learn that in 1938 in UP:

Musammat Pershad had been abandoned by her husband. She then lived as the kept wife of one Ram Chandra of Kotah. But when she became pregnant, Ram Chandra tried to cause abortion. She did not agree. Soon after the birth of the child she was given some medicine by Ram Chandra and she died after 8 hours. Opium was detected in the viscera.[67]

And two years later:

An old man of sixty ... had married a girl of twenty. The girl fell in love with her neighbour, a beautiful young man of her age. One day she was caught by the husband while she was going to her lover's house and severely reprimanded. A few days later the husband was found in an unconscious state. Datura was detected in his vomit.[68]

Such curt and cryptically brief stories of love and betrayal might well have left the reader hankering for more. As with the Fullam-Clark murders, part of the appeal of such stories was the way in which they made the

---

[66] *ToI*, 18 March 1913, 10; ibid., 19 March 1913, 8; ibid., 11 April 1913, 7; ibid., 12 April 1913, 10; ibid., 16 April 1913, 7.
[67] *ARCE (UP), 1938*, 7.   [68] *ARCE (UP), 1940*, 5.

intimate details of home life and sexual conduct matters for public consumption, for edification perhaps but also entertainment.

To take one, much-publicized example, the murder was reported in 1894 of Jadunath Chatterjee, a 60-year-old Brahmin moneylender in the village of Bakshara, not far from Calcutta. At first the crime was attributed to a neighbour, Shama Charan Pal, who was said to have poisoned and then strangled Jadunath following a row over a debt Pal owed to his 'intimate friend'. Not untypically for a genre of colonial writing in which India was represented by obscure but pervasive criminality, this tale of village intrigue and homicidal poisoning was presented to the British public as giving insight into 'the actual facts of the life of the natives' in India.[69] But when the case came to trial at Howrah Sessions Court, Pal was quickly shown to be innocent: instead, Jadunath's wife, Mati Debi, was revealed as the most likely suspect. Ten years earlier, then aged only 25, she had left her husband to go on pilgrimage by herself to Puri in Orissa. This was an unconventional, even defiant, act for a young wife. It was also one of those instances, like the Elokeshi case twenty years earlier, of high-caste women venturing into socially and sexually problematic places outside the home, instances that lay at the heart of many scandals in contemporary Bengal.[70]

At Puri, Mati Debi struck up an intimate (and in this case clearly sexual) acquaintance with Pandu Barik, a low-caste Oriya of her own age, who then returned with her to the house in Bakshara. Ostensibly a servant, Pandu was evidently Mati Debi's live-in lover. Jadunath and his wife no longer ate or slept together, and the liaison between Brahmin wife and low-caste lover was becoming common knowledge locally, so much so that Jadunath was ostracized by fellow Brahmins and barred from using the hookah they normally shared at the village tea shop. Shamed by his wife's infidelity, Jadunath was about to expel Mati Debi and Pandu Barik from his household, and it seems likely that it was the impending expulsion that precipitated the murder. Post-mortem analysis indicated that Jadunath had died from morphine poisoning, not by strangling as originally supposed. However, this could not be conclusively proved as morphine addiction was widespread in the local Brahmin community and so the traces of morphine in Jadunath's body could be attributed to his habitual use of the drug.[71] Police corruption and incompetence, combined with political meddling, further precluded a decisive outcome to

---

[69] Eliza Owen, 'Introduction' to [Anon.], *The Trial of Shama Charan Pal: An Illustration of Village Life in Bengal* (London: Lawrence & Bullen, 1897), vi.

[70] Sarkar, *Hindu Wife*, 81.

[71] Owen, 'Introduction'. For morphine use, see Jadub Kristo Sircar, 'A Case of Morphia Poisoning', *IMG* 14 (1879): 259.

the case, and no sentence was passed. It was still possible, however, for at least one moral to be drawn from this tale – if only for the edification of the British public. 'To the English reader it will seem strange that no law exists in India to regulate the sale of poisonous drugs. If Jadu Chatterjee was poisoned, as the medical evidence suggests, his murderers, whoever they are, could have procured the morphine without difficulty or risk of detection.'[72] It was, by now, a common complaint.

In the Jadunath Chatterjee case, we can again see the difficulty of obtaining conclusive proof for suspected poisoning, whether as a result of poor detective work or because the chemical evidence was elusive and hard to establish as the specific cause of death. But here, too, was a story of multiple intimacies. The companionable intimacy of Jadunath and his friend Shama Charan Pal contrasted with the proscribed intimacies of Mata Debi and her low-caste lover, just as the social intimacy of the village Brahmins, captured by their shared use of the hookah, appeared threatened in Jadunath's case by his wife's ex-marital intimacies. The case, like many other stories of poison, intrigue and murder, revealed intimate details of family life and personal behaviour, from the physical layout of the Chatterjee home, and the telltale proximity of Pandu Barik's mattress to Mati Debi's bed, to the wife's suspicious demeanour and un-widow-like dress on learning of her husband's murder. Within this intricate web of intimacy, poison lost something of its exceptionality (though not its scandalous quality) and became integral to a troubled domestic scene and an unfolding village drama.

Normally the function of practitioners of forensic science was to conduct post-mortems, search for traces of poison in human remains sent for analysis, and indicate the likely cause of death. But in other instances the role of science in relation to poison lay in the perpetration of crime, not in its detection. Harry Clark's use of textbooks of medical jurisprudence and his access, as a doctor, to toxic substances was not altogether unique but was to be found among Indians and Europeans alike. Complicity in a poison plot was one of the accusations made against Dr Calvert in the Bhawal trials. Behind the Pakur murder case of 1932 lurked a fratricidal dispute over control of joint-family property in the Santal Parganas of eastern Bihar. A plot was hatched by Dr Taranath Bhattacharjee, a renowned Calcutta physician, on behalf of Benoyendra Nath Pande, to kill the latter's half-brother Amarendra Pande. The first attempt, worthy of any work of detective fiction, was made using a poisoned pince-nez, rammed down sharply onto the victim's nose, so as to infect him with tetanus. Amarendra became seriously ill but recovered six months later.

[72] Owen, 'Introduction', ix.

In a second, still more elaborate, ruse (one that echoed Harry Clark's quest for cholera bacilli to kill his wife or obscure the effects of arsenic poisoning), Bhattacharjee sent a telegram to the Haffkine Institute in Bombay to request a sample of plague bacillus, claiming (like Clark) that he had discovered a cure for the disease and wanted to test it. The institute refused but, undeterred, Bhattacharjee applied to the city's Arthur Road Hospital for Infectious Diseases, and this time a sample was sent. One day in December 1932, while waiting at Howrah station, Amarendra felt the sharp prick of a hypodermic needle being jammed into his leg. He died soon after. The death certificate, signed by the conspirators, indicated septic pneumonia as the cause of the death, but blood tests carried out in the forensic laboratory in Calcutta revealed plague as the actual cause. Despite a lengthy – and much-publicized – trial from May 1934 to February 1935, too little evidence could be found to convict Benoyendra Nath Pande and Taranath Bhattacharjee for the murder: they were sentenced instead to transportation for life.[73]

Race and other poisons

Let us return once more to the Agra murders. Like many similar cases involving poisoning, the story of the killing of Eddie Fullam and Louisa Clark could be employed discursively as a way of characterizing the social milieu in which they occurred – or the supposedly deviant elements revealed within it. In 1929, soon after retiring as a judge of the Allahabad High Court, Cecil Walsh published his book *The Agra Double Murder*. While claiming to be impartial and objective, his narrative gave singular prominence to the theme of race, almost to the extent that race became the real poison in his story, explaining means and motivation in a way that medical testimony and forensic analysis alone could not. Thus Louisa was described, with some condescension, as coming from 'a good, though humble, Anglo-Indian family'. Her failure to separate herself from Harry, her 'cruel, lustful' husband, was attributed to there being 'much of the fatalism of the East, and of the phlegmatic Eurasian mentality, about her'.[74] Edward Fullam had the virtue of being 'almost pure English'.[75] By contrast, Augusta was seen by Walsh as having betrayed not so much her marriage as her race. As a 'pure' white woman, she had frivolously jettisoned the moral superiority of her race, first by falling for the licentious and duplicitous Clark, and then by conspiring with him to murder her innocent and industrious husband. Worst of all, in this racial narrative, Harry Clark was tainted by being of 'pronounced Eurasian stock', with his

[73] Sarkar, *Notable Indian Trials*, 10–17.   [74] Walsh, *Agra*, 18, 23–25.   [75] Ibid., 27.

dark skin and 'shifty' appearance, and this racial trait appeared, in Walsh's version of the story, to explain his willingness to commit a range of squalid crimes from illegal abortion to arsenic poisoning. 'The heavy stupidity and sensuality of his features' was thus 'reflected in his life and character'.[76]

Walsh went on to explain that, since Clark was born and brought up in India and had lived all his life in 'an Indian atmosphere', he must have been aware of the common confusion that existed in that country between arsenic poisoning and cholera, and so have felt confident of his ability to get away with murder (this was all based on supposition, not on any evidence presented in court). To Walsh it was self-evident that 'Eastern ideas and Eastern thought play a large part in the "make-up" of such a man', adding: 'It is even possible that he inherited from his Hindu ancestry the belief, curious to occidental notions, that murder by poisoning is a less heinous crime than murder by the shedding of blood.'[77] Since Eurasians served as racial intermediaries between whites and 'natives', Clark became the medium through which the moral evil of poisoning, and the physical degeneration caused by 'slow poisoning', was transmitted from East to West.[78] In this context, it was not so much that Eurasians were a 'marginal group' that mattered, as that their proximity to – and intimacy with – 'pure' whites made them a dangerous conduit by which an Indian vice might corrupt and destroy Europeans.[79]

There was little ostensibly racial about the Fullam-Clark murders, though the involvement of Europeans and Eurasians in a sensational murder case undoubtedly heightened public interest just as it caused concern to a race-conscious regime.[80] Augusta Fullam's role in the murders raised awkward issues about the political wisdom of sentencing a white woman (even one convicted of homicide) to death for her crimes. According to one Madras newspaper, 'strong representations' were made to the government about 'the bad effect the hanging of a deliberate poisoner might produce because she was a European, or of European descent, on the respect of the "natives" for persons of the more favoured race and colour'.[81] That Clark, as an officer in the Indian medical

---

[76] Ibid., 19–21.   [77] Ibid., 177–78.

[78] The idea of India as a site of moral and physical 'degradation' among Europeans was long established: see Henry H. Spry, *Modern India, with Illustrations of the Resources and Capabilities of Hindustan* (2 vols, London: Whittaker, 1837), 1: 27–31.

[79] Allen D. Grimshaw, 'The Anglo-Indian Community: The Integration of a Marginal Group', *Journal of Asian Studies* 18 (1959): 227–40.

[80] Whittington-Egan, *Khaki Mischief*, 12–21, casts doubt on several of Walsh's claims about the racial identity of the Clark and Fullam families, suggesting that Augusta and Edward may have been Eurasian, but this is not borne out by contemporary press reports.

[81] *The Commonweal* (Madras), 13 February 1914, 116.

services, also held the military rank of lieutenant, added, from the state's perspective, to the worrying nature of the affair. Indeed, Clark is said to have been 'the only commissioned officer in the Indian army ... ever [to have] suffered the extreme penalty for murder'.[82]

In reality, the Fullam-Clark murder story often seems to obscure rather than to illuminate the race issue. Augusta, European by birth, appeared to have no racial qualms about having a Eurasian, Harry Clark, as her personal physician, entering into an increasingly public relationship with him, and virtually setting up home with him in Agra after her husband's death.[83] Nor does she appear to have been particularly incensed by the racial, as opposed to sexual, connotations of Harry's consorting with 'black hens' in the Meerut bazaar. Instead, the story reveals multiple points of contact and entanglement between Europeans, Eurasians and Indians. The children of both the Fullam and Clark families went to white or mixed-race schools and entered similar careers in the state medical service or the Military Accounts Department. Clark shared homicidal confidences not only with Augusta, but also with his Indian Christian clerk, Alick Joseph, who in addition penned the love letters to Gussie that Harry felt incompetent to write himself. Clark seems to have had no compunction about hiring *budmash*es from the lowest Hindu castes to bludgeon his wife to death in their own home and, as it happened, in front of his own daughter. As servants and employees of one kind or another, Indians slipped in and out of this tale, as they did in so many other colonial poison stories.

What did add to the social and political significance of the Agra killings was the historical moment at which they occurred: 1911, the year of Eddie Fullam's death, was the year in which Eurasians gained official recognition in the census as 'Anglo-Indians'. This designation had hitherto been reserved for India's white residents, giving them a distinctive sense of belonging to India (and of India to them) and so of being more than mere birds of passage. The change was greatly resented by many Europeans as a result.[84] 1911 was also the year in which the Imperial Durbar was held in Delhi for the King-Emperor, George V. An event celebrated with great pomp and elaborate ceremony, the *durbar* was hailed by one Indian physician as 'a time of the greatest

[82] Pearson, *Memories*, 273. Between 1860 and 1947, only four non-military Europeans were executed in India: Kolsky, *Colonial Justice*, 12.

[83] In earlier decades, it had been thought inappropriate for Eurasian doctors to attend European women and their families: Kenneth Ballhatchet, *Race, Sex and Class under the Raj: Imperial Attitudes and Policies and Their Critics, 1793–1905* (London: Weidenfeld & Nicolson, 1980), 102–07.

[84] [*The Times*], *India and the Durbar* (London: Macmillan, 1911), 245.

national rejoicing for the Indian people'.[85] But it did not impinge on the consciousness of Augusta Fullam, who, cocooned in her affair, did not even mention it in her letters. The *durbar* appeared to mark a new zenith in the history of the Raj, to affirm this as an age of unbridled imperial self-confidence and to symbolize the power and permanency of British rule. To its beneficiaries, the Raj seemed at that moment 'as impregnable as a mountain range'.[86]

And yet, even without the Fullam-Clark murders, cracks in the imperial facade were already becoming evident, following sustained opposition to the Partition of Bengal in 1905 (revoked at the *durbar* in December 1911) and by violent acts of revolutionary terrorism.[87] On 17 June 1911, at a time when Eddie Fullam was being fatally dosed with poison, the District Magistrate of Tinnevelly in south India was shot dead as he and his wife sat in a railway carriage. In a further 'outrage', widely reported in the British and Indian press, on 23 December 1912, as the first stage of the Agra murder trials was getting under way, a bomb was thrown at the Viceroy during his official entry into Delhi, the new imperial capital. The constitutional concessions made by the British through the Indian Councils Act of 1909, themselves evidence of a need felt by the British to compromise with nationalist demands, had only partially succeeded in assuaging the quest for self-government and ultimately for freedom from colonial rule.[88]

Against such a background, the Fullam-Clark murders might be read not so much as an isolated, aberrant episode as a further sign of the waning power and faltering prestige of the British Raj. Indeed, for reasons examined more fully in the next chapter, the years surrounding the Agra murders, between about 1890 and 1914, marked a moment when poison and poisoning took on a new administrative and political significance in India. The period was one in which long-proposed measures to control the sale of poisons finally became law in 1904. Apart from individual acts of poisoning that made for attention-grabbing headlines, like the Fullam-Clark case, numerous other poison cases, affecting Indians as well as Europeans, came to light at this time. It was as if the political atmosphere of India had itself taken on a new toxicity. It was surely no coincidence that in this period of mounting imperial anxiety, a flurry of books

---

[85] S. M. Varis, *A Study of Malaria and Beri-Beri* (Allahabad: Pioneer Press, 1912), i.

[86] Peter Greave, *The Seventh Gate* (Harmondsworth: Penguin, 1978), 10.

[87] Fifty years on from 1857, the British feared renewed Indian unrest: Kim A. Wagner, *The Great Fear of 1857: Rumours, Conspiracies and the Making of the Indian Uprising* (Oxford: Peter Lang, 201), xv–xxiii.

[88] For these developments, see Sumit Sarkar, *Modern India, 1885–1947* (2nd ed., Basingstoke: Macmillan, 1989), 137–46.

appeared in Britain reminding the public of the nature and extent of Indian criminality – from *thugi* to child murder – and the existence of poison 'outrages' – from the Baroda case to the Agra murders – that could be seen in one way or another to typify India and demonstrate its inherent unfitness for self-rule.[89] In this sense, poison had not one public but two that perhaps only marginally overlapped. In addition to the mostly middle-class Indians who read about poison trials and murder cases, and who felt their lives to be threatened by poison in one form or another, there was also a wider imperial public – the readers of newspapers and of books about Indian history, society and politics in Britain, across the Empire and even further afield, in Continental Europe and the United States.

In 1930, a year after publishing his book on the Agra murders, Cecil Walsh produced a more general – but still more politically and racially charged – account of crime in India. The Agra case was again mentioned in passing – this time to indicate how often women were implicated in domestic poisonings – but Walsh's primary concern was with crimes committed by Indians against other Indians.[90] In the introduction to this second book, Walsh explained that his aim was to offer his readers a collection of tales of poisoning and other crimes, and so

to present some sort of picture of the mentality, the duplicity and cunning, the indifference to human life, the callous indulgence in false evidence and false charges and the lack of moral fibre which daily manifest themselves among the millions of cultivators whom we govern, and of whom the Englishman at home knows so little.[91]

This might almost have been Norman Chevers in the 1850s and 1860s. But, writing in 1930, at a time when the nationalist movement under Gandhi was winning mass support, Walsh had a more contemporary moral in mind: left alone, the ordinary Indian villager was generally law-abiding, but was 'easily led and easily misled, and is dangerous in crowds if aroused'. Governed by 'tradition and instinct', Indians lacked 'that public opinion which is the chief moral force amongst civilized people, and the foundation both of Society and of Public Order'.[92] Poisoning was political.

Poisoning in a literal sense was never employed as a deliberate weapon in India's nationalist struggle – though (as seen in Chapter 3) it could be

---

[89] Including H. L. Adam, *The Indian Criminal* (London: John Milne, 1909); Edmund Cox, *Police and Crime in India* (London: Stanley Paul, 1911); S. M. Edwardes, *Crime in India* (London: Oxford University Press, 1924); Augustus Somerville, *Crime and Religious Beliefs in India* (Calcutta: The Criminologist, 1929); J. C. Curry, *The Indian Police* (London: Faber & Faber, 1932).
[90] Cecil Walsh, *Crime in India* (London: Ernest Benn, 1930), 173.   [91] Ibid., 9.
[92] Ibid., 10.

assigned a role in the resistance of Indian princes to overbearing Residents. But that did not prevent at least a few European officials from believing that they had been poisoned for political reasons. There was, for instance, the case of J. A. Ferguson of the Punjab Civil Service, who in 1923 claimed that he had been poisoned with arsenic in retaliation for his part in combating nationalist agitation in the district of Rohtak, close to Delhi. In making his case to the India Office in London (which at first thought his claims far-fetched), Ferguson explained his personal sense of vulnerability by observing: 'It was easy for anyone to poison me. My sugar & milk were bought daily in the bazaar; there were often strangers in my kitchen, any one of whom could have slipped arsenic into the milk jug before tea time.'[93] Despite initial scepticism, officials in London began to believe that the medical evidence might, after all, support Ferguson's claim, but, as so often the case, no conclusive proof of poisoning could be found. Having recovered his health in England, Ferguson returned to his administrative duties in India.[94]

Hard evidence might be difficult to come by, but the possibility of poisoning as an insidious, secretive, subversive act fuelled colonial fears and racial paranoia. It was, by its very elusiveness and its ubiquity (real or imagined), one of the uncertain 'terrors' that beset British rule.[95] It was one of the many lurking, but indeterminate, factors that ensured that the British could never quite feel 'at home in empire'.[96] In a figurative, rather than material, sense, poison helped define, and give rhetorical force to, what was seen to be the treacherous nature of the opposition the British found themselves confronting. India seemed, to many Westerners, to be entering an era of 'race-hatred and race-contempt'. It was indicative of this darkened mood that one of the most militant and vociferous opponents of British rule, the revolutionary Shyamji Krishnavarma, was said to be spreading 'the poison of hate and murder' against Europeans.[97]

Poisoning lent itself to a wide variety of poison stories and toxic tales. Some of these used the poison motif to suggest a dangerous (if unavoidable) intimacy between colonial rulers, their servants and subjects; others communicated a collective anxiety felt by colonialists in an alien and, in some respects, increasingly hostile land. Such stories, blending fact with supposition, helped fuel the colonial imagination. But, often quite

---

[93] J. A. Ferguson to Sir Malcolm Seton, India Office, 25 April 1923, L/P&J/6/1849: 2645, IOR.

[94] J. A. Ferguson to Sir Malcom Seton, India Office, L/P&J/6/1849: 2645, IOR.

[95] Suleri, *Rhetoric*, 2–4.

[96] Ranajit Guha, 'Not at Home in Empire', in Partha Chatterjee (ed.), *Ranajit Guha: The Small Voice of History: Collected Essays* (Ranikhet: Permanent Black, 2009), 441–54.

[97] J. N. Farquhar, *Modern Religious Movements in India* (New York: Macmillan, 1915), 356, 359.

consciously, they were also replicated and embellished to educate an imperial public about the dangers – to Britain but also to India itself – that persisted in that country and imperilled its future, even after decades of British rule. But there were many poison tales that did not depend on European anxieties, on imperial vanity and self-vindication, for their moral message. They highlighted episodes of conflict within Indian society, for the amusement or edification of Indians themselves. They touched on issues of decency and impropriety, danger and deceit, especially in relation to women, marriage, property and the intimacies of the home. Here were issues that had meaning and significance for India's own population, especially, but never exclusively, for Indians of the middle classes. In this regard, and, as will be seen in the following chapters, in other respects, too, poison was a potent metaphor and a powerful material presence around which new notions of public and private life began to congeal.

In all of this, science played a part – in the detection, and even sometimes in the commission, of poison plots. But much-publicized trials and lurid newspaper reports also gave science a wider audience, as medical experts and forensic specialists strove – not always convincingly – to present their evidence in court and uphold the credibility of their findings. By the early decades of the twentieth century, poisons and poisoning could no longer be ignored. Not only had they become a matter of state concern, but they were also too much in the public eye to be denied. In one form or another, as danger or as opportunity, toxicity had now to be embraced.

6   Embracing toxicity

At the close of the nineteenth century and into the early years of the twentieth, the issue of poisons and poisoning took on a new character in India. Although the use of poison to commit murder or suicide continued, and included high-profile cases like the Agra murders of 1911–12, the focus of public interest and state concern about poisoning began to shift elsewhere. It moved increasingly to forms of poisoning associated with urban living and the urban marketplace (such as the adulteration of foodstuffs and the impact of industrial pollution), and, in the countryside, to the killing of cattle. Instead of vegetable drugs like aconite, datura, opium and nux vomica, which had dominated colonial thinking about poisonous drugs and dangerous medicines for much of the nineteenth century, arsenic increasingly came to be thought of in India (as elsewhere) as the greatest single danger to human life. This widening concern found expression in the Indian Poisons Act of 1904, passed decades after similar enactments in Britain.

But the changing engagement with toxicity took other forms too. As indicated in previous chapters, Indians, especially those working for the chemical examiners' departments, played a salient role in the investigation of toxic drugs and, in response to the concerns of the Indian public as well as those of the colonial state, pressed for tighter controls over the sale of arsenic and other poisons. In other respects, too, an expanded notion of poisons and poisoning became a matter Indians engaged with as critically important to their own lives and livelihoods. By the 1890s and 1900s, the rise of germ theory and bacteriology and the growing sophistication of chemistry and biochemistry gave toxicology a new scientific stature and increased importance in policing poisons in the home, the bazaars and the environment at large. This was not just laboratory science, important though the laboratory had become as a site for the performance of science. Toxicology also served a very public function: to inform, educate and reassure. The old stereotypes of Indian poisoners – thugs, princes, women – did not disappear overnight, but they gradually receded from colonial rhetoric and public discourse, to be replaced by a new cast of poisoners and polluters. If in legend 'poison maidens' once symbolized a toxic embrace to be shunned and feared, by the early and middle decades of the twentieth century toxicity appeared more as a principle requiring active engagement – a danger to be overcome but also, where practicable, utilized and embraced for constructive ends. Toxicology – a science predicated on adversity – could be turned to positive effect, and, from thinking like a colonial state, toxic science could turn to thinking like a nation.

The arsenical empire

Arsenic was critical to this unfolding story. White arsenic (arsenious oxide) was not a commonly occurring mineral in India. More common were yellow orpiment and red realgar, naturally occurring sulphides of arsenic, which were to be found in Himalayan regions like Chitral and Kumaon or, in India's long history of metallurgy, were a familiar by-product of smelting copper, tin, lead and other ores.[1] In the early nine-teenth century, arsenic was one of many curative commodities traded, in one direction or the other, between Britain and India, alongside aloes, arrowroot, castor oil, nux vomica and musk.[2] In the second half of the century, the bulk of India's arsenic either came overland from China or was imported by sea from Europe. On a smaller and less dramatic scale than opium, arsenic, too, had by then become a global commodity. Indeed, India's 'arsenic century' (to borrow James Whorton's phrase) demonstrates the paradox of an imperial connection in which the Empire was both the primary commercial source of a toxic substance and the political agency that, ultimately, sought its containment and control.[3] Arsenic represents the global in this toxic history: poison as an object of trans-regional trade and capitalist enterprise, poison as the subject of criminal practices and systems of governance that spanned empires and crossed continents.

The uses of arsenic in nineteenth-century India were not unlike those in contemporary Britain. In 1904, Bombay's Chemical Analyser listed twenty-two different uses for arsenic.[4] White arsenic and orpiment were employed in a variety of industrial processes, including the preparation of

---

[1] J. Coggin Brown, *Notes on Antimony, Arsenic and Bismuth* (Calcutta: Superintendent, Government Printing, India, 1921), 15–16.

[2] Madras Dispatches, 28 October 1814, E/4/914, IOR.

[3] James C. Whorton, *The Arsenic Century: How Victorian Britain Was Poisoned at Home, Work, and Play* (Oxford: Oxford University Press, 2010).

[4] Collis Barry, *Legal Medicine (in India) and Toxicology* (2nd ed., 2 vols, Bombay: Thacker, 1904), 1: 368.

leather, the manufacture of paints, dyes and whitewash, textile production, the jewellery trade and paper making. In 1843, as Chemical Examiner for Bengal, Frederic Mouat expressed alarm at the unregulated sale of arsenic and other poisons in the Indian bazaar, but, a dozen years later, in 1855, as Inspector of Prisons, he recommended the use of white arsenic to improve the colour and quality of the paper made by prisoners. He did, though, admit to some misgivings when he found orpiment being 'carelessly handled' in one jail, fearing that it might 'lead to mischief in the hands of any designing or vindictive prisoner'.[5] As a taste enhancer and preservative, arsenic entered many items of food and drink. One illustration of this was the imported beer drunk by Europeans and Eurasians, a lifestyle factor that helped explain higher arsenic levels in their urine and faeces than in most Indians.[6] Arsenic was present in the brightly coloured sweets and painted toys sold in the bazaars, exposing children to low, but occasionally fatal, doses, and it was used medicinally by *vaids* and hakims as well as by practitioners of Western medicine. Arsenic could be found in the Indian or European home in flypapers, rat poison, insecticides, tonics, 'love philtres' and abortifacients. Mixed with quicklime, it was widely used as a depilatory. Arsenic was valued as much for what it preserved as for what it removed. Taxidermists kept arsenic to preserve their specimens. When Fanny Parks trapped a large scorpion in her bathroom, she preserved it with arsenical soap before adding it to her private museum.[7] As in Europe, there was a long-standing belief that arsenic in sub-toxic doses improved the condition of horses, gave them a glossy coat and could cure foot canker.[8] There were also arsenic eaters in India, who, like the Styrians in Austria, believed it conducive to sexual prowess, good health and longevity.[9]

And, as we have seen previously, arsenic became the poison of choice for murderers or would-be assassins – in the 1874 Baroda incident, in the Wagner-Cray case of 1895, in the Agra murders of 1911–12 and in a large percentage of the suspected poison cases referred to the chemical examiners for analysis. As an instrument of murder, arsenic was at the height of

---

[5] Fred. J. Mouat, *Reports on Jails Visited and Inspected in Bengal, Behar, and Arracan* (Calcutta: Military Orphan Press, 1856), 21, 33, 159.

[6] [Anon.], 'Current Topics', *IMG* 36 (1901): 71; K. N. Bagchi and H. D. Ganguly, 'Arsenic in Human Tissues and Excreta', *IMG* 72 (1937): 479.

[7] Fanny Parks, *Wanderings of a Pilgrim in Search of the Picturesque* (2 vols, London: Pelham Richardson, 1850), 1: 61.

[8] J. D. E. Holmes, *A Note on Some Interesting Results Following an Internal Administration of Arsenic in Canker and Other Diseases of the Foot in Horses* (Calcutta: Superintendent, Government Printing, India, 1912).

[9] Gopal Chandra Mukherji to Deputy Commissioner, Manbhum, 7 August 1899, India, Home (Judicial), no. 210, March 1900, IOR.

its popularity between the 1860s and 1920s. Of the 4,719 suspected poisonings investigated by the Chemical Examiner in Punjab between 1861–66 and 1879–87, 1,286 were ascribed to arsenic and only 350 to opium and 180 to datura.[10] In Bombay, between 1875 and 1884, 507 of the 947 deaths attributed to poison were due to arsenic compared to 151 from opium and 74 from datura.[11] In one of many such cases reported in the *Indian Medical Gazette*, in 1873, a Punjab surgeon recounted how he had supplied two ounces of arsenic to a hospital assistant to kill rats. The assistant, however, used it to poison Goolab, the *havildar* with whose wife he was having an affair. Twelve days after the poisoning began, Goolab began to vomit and complained of a 'gnawing hot pain' in his stomach. He thought his food had 'disagreed with him' but said he now felt 'all right'. He died soon after. The surgeon insisted that a post-mortem be held, even though this was 'very obnoxious' to the deceased's friends, and traces of arsenic were found in Goolab's body. The wife was found guilty and sentenced to transportation, but no evidence could be found to incriminate the hospital assistant.[12]

For the 'ignorant and unscientific poisoner', the appeal of white arsenic was obvious: it was odourless, was tasteless (or, according to Colonel Phayre in Baroda, had a metallic taste), and dissolved easily in liquid or when mixed with food.[13] White arsenic was cheap to buy and 'could be obtained in every native bazaar'.[14] Given its many domestic uses, in India possession of arsenic did not necessarily afford 'a strong presumption of guilt' in criminal cases.[15] Conversely, however, the presence of arsenic was relatively easy to detect through the well-established Marsh test, but also from the thick yellow deposit found on internal organs in many Indian cases as a result of the very large quantity of arsenic administered.[16] But not all arsenic poisoning was intentional. At home or in the marketplace, the unwary could easily confuse white arsenic

---

[10] T. E. B. Brown, *Punjab Poisons: Being a Description of the Poisons Principally Used in the Punjab* (3rd ed., Lahore: 'Civil and Military Gazette' Press, 1888), 19.

[11] [Anon.], 'Poisoning in India', *British Medical Journal*, 17 September 1892, 642.

[12] J. W. Johnston, 'Poisoning by Repeated Small Doses of Arsenic', *IMG* 8 (1873): 185–86. On the symptoms of arsenic poisoning, see Katherine Watson, *Poisoned Lives: English Poisoners and Their Victims* (London: Hambledon Continuum, 2004), 4–15.

[13] Robert Harvey, 'Report on the Medico-Legal Returns Received from the Civil Surgeons in the Bengal Presidency during the Years 1870, 1871, and 1872', *IMG* 11 (1876): 61.

[14] V. Ball, *A Manual of the Geology of India, Part III: Economic Geology* (Calcutta: Office of the Geological Survey of India, 1881), 162.

[15] C. H. Bedford, 'Notes on Some Toxicological Experiences in Bengal and in the Punjab', *IMG* 37 (1902): 203.

[16] W. J. Buchanan, 'A Chapter on Medical Jurisprudence in India', in Fred. J. Smith (ed.), *Taylor's Principles and Practice of Medical Jurisprudence* (7th ed., London: J. & A. Churchill, 1920), 2: 912.

crystals with sugar or salt. In 1899, L. A. Waddell, Chemical Examiner for Bengal, observed: 'Nearly all Muhammadans keep arsenic in their houses as a depilatory, and the poison is sold in the bazar by the same shop-keepers who sell spices and salt. The risk therefore of this deadly poison finding its way accidentally into the food is very great.'[17]

In view of its wide spectrum of uses, and its relative cheapness, it is not surprising that imports of arsenic grew substantially as Indian trade and industry expanded. In 1878–79, 110 tons of arsenic were imported, with a value of Rs 38,720: this was up from 57 tons in 1877–78, and 27 tons in 1876–77.[18] The likely explanation for this fourfold rise in two years was the surge in the leather trade in India during the 1876–78 famine, in which huge numbers of cattle died or were sold by peasants solely for the value of their hides.[19] Although the true extent of its use for this purpose is unclear, arsenic was reputedly 'one of the cheapest and most effective preservatives of leather and skins'.[20] By 1913–14, a decade after the Indian Poisons Act – India's arsenic act – had come into force, imports into India stood slightly higher than in 1878–79, at 147 tons, with a value of £4,000.[21] This was not a notably large or particularly lucrative import trade. Its real commercial significance lay in arsenic's close connection with the leather industry and the preservation of cattle hides that were sold to the United States and Europe. In the 1880s and 1890s, more than 7 million hides a year passed through Calcutta, the main hub of this trade: in 1900–01 alone, hides and skins worth nearly Rs 113 million were exported.[22]

Like the controversy over opium, arsenic poisoning became an international issue. In Britain, alarm over accidental poisoning in food, drink and medicines, and the trials of male and (more especially) female arsenic murderers, provoked an 'arsenic panic' in the 1840s. Laissez-faire objections were overruled in favour of state regulation of the sale of arsenic in 1851, followed in 1852 and 1868 by two Pharmacy Acts to control the public availability of this and other poisons.[23] Arsenic was categorized in India, too, as a homicidal poison, but when the issue of regulation was

---

[17] *ARCED (Bengal), 1899*, 13.   [18] Ball, *Manual*, 572–73.

[19] Alfred Chatterton, *Monograph on Tanning and Working in Leather in the Madras Presidency* (Madras: Superintendent, Government Press, 1905), 6.

[20] Horace A. Cockerell, Secretary, Bengal, Judicial, Political and Appointments, to Secretary, India, Home, Revenue and Agricultural, 18 June 1880, Bengal, Judicial, no. 42, June 1880, IOR.

[21] Brown, *Notes*, 18.

[22] Chatterton, *Monograph*, 11; E. H. C. Walsh, Secretary, Board of Revenue, to Secretary, Bengal, Municipal (Medical), 4 February 1896, India, Home (Judicial), nos 444–508, May 1899, IOR.

[23] Whorton, *Arsenic Century*; Watson, *Poisoned Lives*, 42–43, 58.

raised with the Government of India in the 1850s, the proposal was dismissed on the grounds that restricting the sale of arsenic could only be justified 'by the prevalence of a known evil arising from it in unusual extent, and by the impossibility of finding other means for its repression'.[24]

One of the complicating issues with regard to arsenic (as with opium) was that it was seen by many allopathic doctors, as well as practitioners of indigenous medicine, as an invaluable therapeutic, more medicine than poison. In the form of 'Tanjore pills', arsenic was adopted in the late eighteenth century as a cure for snakebite by Company surgeons in south India influenced by regional Siddha medicine. Its therapeutic use spread to Britain, one physician observing in 1787 that, although arsenic was 'strongly connected with the idea of poison', the time had come to accept it as a 'safe, active and effective medicine'.[25] The spate of murders and accidental deaths in subsequent decades in Britain produced a strong reaction against all but the most cautious medicinal use of the drug, but in India the reaction was by no means so marked. Like opium, arsenic had its supporters as well as sceptics. India's diseases – with cholera and malaria pre-eminent among them – appeared so debilitating and deadly that only a drug as powerful as arsenic seemed to have the capacity to check them or assuage their effects. The fact that both cholera and malaria were themselves widely referred to as 'poisons' made it the more compelling to think of the need for another poison – like arsenic – to counteract them.[26] Prophylactic injections of quinine and arsenic were recommended to protect indentured labourers against cholera on voyages from Calcutta to the West Indies in the 1860s and early 1870s.[27] In the early 1890s, arsenic – rather than quinine – was still widely touted as a prophylactic and antiperiodic in the treatment of malaria.[28]

In 1916, George Dennys of the IMS questioned the value of quinine in the treatment of this disease and claimed instead to have successfully used – on both Indian and European patients – a combination of white arsenic and iron citrate as a prophylactic and cure for 'chronic malarial poisoning'. He acknowledged that white arsenic was generally thought of

---

[24] Cited in *RCE (Punjab), 1873*, 15.
[25] Pratik Chakrabarti, *Materials and Medicine: Trade, Conquest and Therapeutics in the Eighteenth Century* (Manchester: Manchester University Press, 2010), 186.
[26] On cholera and malaria 'poisons', see G. C. Roy, 'Observations on the Nature of Cholera Poison', *IMG* 8 (1873): 120–22; J. Fayrer, 'A Case of Acute Malarial Poisoning', *IMG* 26 (1891): 296–301.
[27] William H. Pearse, 'The Prophylactic Use of Arsenic and Quinine against Cholera', *IMG* 17 (1882): 190–91.
[28] W. B. Bannerman, 'Note on Arsenic as a Prophylactic for Malaria', *IMG* 26 (1891): 70–71.

as a 'virulent poison', which no ordinary person could endure even in small doses without suffering acute gastric irritation.[29] However, his experience showed that it was possible gradually to increase the quantity of arsenic administered – orally or by injection – until, after three or four months, the patient was able to tolerate a relatively large dose. Dennys continued: 'The terrors of chronic arsenical poisoning described in books on therapeutics have never once come to my notice, though many of my patients have continued the treatment for several months.' The only 'disagreeable symptoms' he knew of were 'colicky pains with a tendency to looseness of the bowels, a heavy and sometimes hot sensation in the stomach, and very occasionally a wateriness and irritation of the eyelids'.[30] But while many allopathic practitioners continued to favour the medicinal use of arsenic, others, associating it with the malpractice of *vaid*s, hakims and 'quacks', saw it as a poison, causing prolonged sickness, paralysis and death.[31] In 1856, James Ranald Martin described arsenic as a powerful tonic and antiperiodic, used 'from time immemorial' by the 'native doctors' of India. But, he cautioned: 'I saw at the Native Hospital of Calcutta daily examples of permanent injury to the circulating system and to the mucous digestive organs from the abuse of this mineral by the Bazar practitioners, in their endeavours to cure fevers, rheumatism, and venereal diseases.' His advice was that it should be used with extreme caution and only in the treatment of intermittent fever.[32]

Responding to local concerns as well as metropolitan precedents, in 1866, the Government of Bombay introduced legislation to 'regulate and restrict' the sale of 'poisonous drugs and deleterious substances'. The act's schedule listed several plant poisons, including aconite, datura and nux vomica, but also mineral poisons, most of which were forms of arsenic – white arsenic, orpiment, realgar, Scheele's green and Schweinfurt green, the last two arsenic compounds used in dyeing. Under the act, modelled on earlier legislation in Britain, the proscribed poisons could only be sold by licensed vendors and not to anyone under the age of 18. All sales were to be recorded in English or the vernacular, and poisons could not be sold to a purchaser unknown to the vendor without a witness testifying to his or her identity. White arsenic, the main target for this legislation, was only to be sold mixed with soot, indigo or

---

[29] George W. P. Dennys, 'Iron and Arsenic as a Cure for and a Prophylactic against Malaria', *IMG* 51 (1916): 242.

[30] Ibid., 243.

[31] For example, T. Hume, 'Case of Partial Paralysis, Supposed to Have Followed the Injudicious Administration of Arsenic', *IMG* 11 (1876): 103; *ARCED (Bengal), 1900*, 8, 10; *ARCED (Bengal), 1903*, 5.

[32] James Ranald Martin, *The Influence of Tropical Climates on European Constitutions* (London: Churchill, 1856), 167–68.

Prussian blue so as to prevent mistaken use.[33] However, the act proved virtually inoperable, even in Bombay city, not only from the lack of any dedicated agency to check sales and inspect registers but also because, in order not to penalize allopathic and indigenous physicians, the legislation had excluded the sale of any of the listed drugs when used as medicines. The requirement that arsenic be dyed blue was seldom complied with in India where white arsenic tended to be sold in lumps, not as a powder that could be coloured.[34] In consequence, the criminal use of arsenic continued unchecked. In Bombay city in 1870–71, twenty-three deaths were attributed to arsenic poisoning (there were a further fifteen such deaths in 1871–72), in addition to numerous arsenic cattle poisonings.[35] The clear deficiencies in the Bombay act discouraged other provinces from following suit.

Cattle poisoning

One of the factors that eventually forced a change of mind on the part of government was the reported rise of cattle poisoning by means of arsenic. It may seem odd that cattle poisoning should occur – on an apparently substantial scale – in a country where a large proportion of the population revered the cow and regarded its killing with deep abhorrence. In the 1880s, in an effort to reassert Hindu religious identity and mobilize popular support, a 'cow protection movement' was formed in northern India to save animals from slaughter at the hands of Muslim butchers and beef-eating Britons.[36] Cattle poisoning was not the cause of this movement, though the religious significance attached to the cow was one factor in explaining why the British and many high-caste Indians regarded the crime with particular repugnance. But cattle were also property, among the few possessions poor cultivators could lay claim to: they were working animals, essential for ploughing, drawing water and other agricultural tasks. And, as we have seen, cattle hides were a commodity of increasing value as India's leather industry and export trade burgeoned in the second half of the nineteenth century. Cattle killing and datura *thugi* were near-contemporaries, both originating in the 1850s. They affected the same regions of northern and central India and involved many of the same

---

[33] *Bombay Acts, 1862–70*, V/8/180, IOR.
[34] See 'Reports Regarding the Working of Bombay Act VII of 1866', in *Report of the Commissioners Appointed to Inquire into the Origin, Nature Etc., of Indian Cattle Plagues* (Calcutta: Office of the Superintendent of Government Printing, 1871), 738–43.
[35] *RCA (Bombay), 1871–72*, 5.
[36] Gyanendra Pandey, *The Construction of Communalism in Colonial North India* (Delhi: Oxford University Press, 1990), 164–75.

issues of evidence, identity and governance. And yet the rise of cattle poisoning seemed more alarming to colonial officialdom than the allegedly systematic use of datura in road robbery and had more momentous results.

Two forms of cattle poisoning were thought to exist in India. One involved the use of a poisonous spike or needle, known as a *sui* or *sutari*, which was occasionally also used in homicides. This was made from the seeds of *Abrus precatorius*: these could be purchased in Indian bazaars, thus adding to the long list of widely available Indian plant poisons.[37] The seeds were softened, pounded into a paste, then rolled into a point which, on hardening, became as sharp as steel. The spike was thrust into the head of the animal, close to the brain, into its neck or hindquarters, causing death within forty-eight hours.[38] A more common practice involved the use of a poison, usually arsenic, mixed with fodder or sugar. In the Bombay Presidency, an ear of *jowar* millet was hollowed out and filled with up to two ounces of arsenic: the outer part of the ear was then tied or sewn back before being given to cattle to eat.[39] Once ingested, arsenic had much the same effect on cattle as on humans. Sub-toxic doses of arsenic might improve a horse's coat but it did not require much arsenic to kill a cow.

In both techniques the logic was presumed to be the same. For high-caste Hindus touching a dead cow or working with leather was ritually polluting. By custom, cows that died a natural death became the property of untouchables, who removed the dead animals and stripped the carcass of its hide. From this were made buckets to raise irrigation water, sandals, belts and other household or agricultural goods. In northern and central India, the untouchable community mainly identified with removing dead animals (and, according to colonial ethnography, with leather working) were the Chamars. What apparently happened from the 1850s onwards was that Chamars were deliberately poisoning cattle rather than waiting for them to die naturally. They then sold the hides to merchants, mostly Muslims based in Calcutta and Patna, who had previously supplied them with arsenic for the purpose. The expansion of the Indian leather industry and the flourishing export of cattle hides from the mid-1870s gave a further impetus for cattle killing, though this explanation need not

---

[37] Norman S. Rudolf, 'Notes on the Chemistry of the Seeds of *Abrus precatorius* and "Sutari" Poisoning', in *Transactions of the First Indian Medical Congress, Calcutta* (Calcutta: Caledonian Steam Printing Works, 1895), 487–88; G. D. Mall, ' "Sui" or Needle Poisoning in the Punjab in Cattle', ibid., 503–05.

[38] H. M. Ramsay, *Detective Footprints: Bengal, 1874–1881* (London: Army and Navy Co-Operative Society, 1882), 44–55; V. R. Gopalakrishnan, 'Cattle Poisoning in Assam', *Indian Farming* 5 (1944): 77–79.

[39] *RCA (Bombay), 1871–72*, 3.

preclude the possibility that cattle were sometimes attacked purely from spite or from motives of revenge.[40]

The first substantial report of cattle poisoning came to the attention of the authorities in the North-Western Provinces in 1854. A large traffic in white arsenic was reported by George Campbell, the District Magistrate of Azimgarh, which could not be explained by medicinal need or any other legitimate use. Campbell's suspicions fell on the wholesale traders reputedly supplying Chamars with arsenic – which of late, he reported, had become 'cheap, plentiful, and common' – in order to poison cattle for the sake of their hides. One Patna merchant alone had imported four tons of arsenic from Calcutta over the past year, and there appeared no plausible reason for such large quantities of 'a most deadly poisonous drug' unless to kill either people or cattle.[41] Campbell's report met with some scepticism at the time, as it has done since. Indeed, Saurabh Mishra has argued that the Azimgarh magistrate, inspired by Sleeman's suppression of *thugi*, effectively invented a myth of cattle poisoning as an 'oriental crime', and concocted or extracted by means of coercion evidence to incriminate Chamars. Once the idea of Chamars as cattle-killing criminals had become established in the 1850s, later reports in the 1870s and 1880s further cemented the stereotype of Chamars as a community of low-status poisoners and leather workers (rather than as agricultural labourers unconnected to the leather trade).[42]

If this analysis is correct, then the story of cattle poisoning suggests many similarities with Colonel Hervey and datura *thugi*. The claims made by Campbell for systematic cattle poisoning recall Hervey's campaign in several respects – in the continuing reference back to the authoritative example of *thugi* as a characteristically Indian mode of widespread, organized crime, in the way in which poisoning (if here with arsenic rather than datura) served as a central trope in these latter versions of Indian criminality and in the way in which ethnic and communal stereotypes gained self-perpetuating authority and might even outweigh the apparent paucity of evidence. That said, though, it would still be necessary to explain why the chemical examiners, using one of their more dependable techniques of detection, saw (or believed they saw) material evidence for the poisoning of cattle – and that in considerable numbers – and why the

[40] *RCA (Bombay), 1947*, 3. *Sui* poisoning, as well as poisoning by arsenic, was said to be the special preserve of Chamars seeking cheap hides for their leather working: L. A. Waddell, *Lyon's Medical Jurisprudence for India* (5th ed., Calcutta: Thacker, Spink, 1914), 29, 568.

[41] G. Campbell to Superintendent of Police, 5th Division, Benares, 16 November 1854, in *Selections from the Records of Government, North Western Provinces, Vol. 5* (Agra: Secundra Orphan Press, 1856), 276–77.

[42] Saurabh Mishra, *Beastly Encounters of the Raj: Livelihoods, Livestock and Veterinary Health in North India, 1790–1920* (Manchester: Manchester University Press, 2015), ch. 6.

government in India came to see cattle poisoning as a more substantial danger than the datura *thugi* on which Hervey, as Superintendent of the Thugi and Dacoity Department, a more prominent figure than Campbell, staked his professional reputation.[43]

Campbell's initial claims met with some disbelief. In the wake of his 1854 report and in response to enquiries from Calcutta, Patna's District Magistrate could offer no alternative explanation for the increase in local arsenic imports and yet remained doubtful about its use to kill cattle, remarking that such an act would be 'looked upon by Hindoos generally as a heinous offence against their religion'.[44] No further action was taken at the time, and the region soon became engulfed in the 1857 uprising. But once the revolt was over, the connection between Chamars, cattle poisoning and the leather trade was quickly re-established. In his treatise on medical jurisprudence, Norman Chevers drew renewed attention to the seemingly widespread crime of cattle poisoning and linked it directly to rising arsenic imports.[45] An act to prevent the 'Malicious or Wanton Destruction of Cattle' had been introduced in 1856, and though it was soon after superseded by the Indian Penal Code, the number of suspected cattle poisoning cases continued to rise. Between 1865 and 1869, 1,462 cases were prosecuted in the courts of northern and central India: most involved arsenic poisoning by Chamars.[46] Indeed, the crime was said to be 'practically confined to the Chamar caste', and, while in other contexts possession of arsenic was not seen to constitute evidence of criminal intent, in cases involving Chamars their having white arsenic was taken as sufficient proof of criminal involvement.[47]

The task of proving that cattle had been poisoned was ultimately entrusted, as with human poison cases, to the provincial chemical examiners, and by the late nineteenth century the examination of animal

---

[43] One difficulty with Mishra's analysis is that he treats the science of the period with undue scepticism, referring to toxicology as a 'science' only in inverted commas, when its scientific methodology was well established, and doubting the value of the Marsh test for arsenic when its practical use was widely established internationally by the 1880s and already had a long history in India going back to O'Shaughnessy in the 1830s: ibid., 133–36.

[44] W. Ainslie to Secretary, Bengal, Judicial, 31 January 1855, *Selections from the Records of Government, North Western Provinces, Vol. 5*, 281.

[45] Norman Chevers, *A Manual of Medical Jurisprudence for India* (3rd ed., Calcutta: Thacker, Spink, 1870), 132; Saurabh Mishra, 'Of Poisoners, Tanners and the British Raj: Redefining Chamar Identity in Colonial North India, 1850–90', *Indian Economic and Social History Review* 48 (2011): 331–32.

[46] 'Report on Cattle Poisoning', in *Report of the Commissioners Appointed to Inquire into the Origin of Indian Cattle Plagues*, 645.

[47] Eustace J. Kitts, *Serious Crime in an Indian Province* (Bombay: Education Society's Press, 1889), 60; R. D. Spedding to Commissioner, Benares Division, 11 October 1879, Bengal Judicial, nos 38–39, June 1880, IOR.

viscera had become a significant part of their work. In NWP, between 1884 and 1894, the annual number of cattle poisoning cases taken up for investigation ranged between 67 and 102, while human cases stood at around 250 a year. By 1900, as cattle killings reached their peak, the number of suspicious cattle deaths had reached 163, though human cases examined had risen still further to 258.[48] In Bengal, in 1899, the Chemical Examiner assessed 380 human and 148 cattle cases, and while among humans opium accounted for 27 per cent of deaths investigated, white arsenic and orpiment for 12 per cent and aconite for 5 per cent, 75 per cent of animal cases were attributed to arsenic.[49] In the Bombay Presidency, between 1875 and 1884, 743 cases of cattle poisoning were examined, of which all but 18 were identified as due to arsenic.[50] Five years further on, in 1891, Bombay's Chemical Analyser was asked to investigate 151 cases of poisoning in humans but more than twice that number – 321 – in cattle: by 1908, the ratio had reversed with 123 suspicious cattle deaths compared to 281 human cases.[51] By 1921, the number of human cases referred to Bombay's Analyser far outweighed those relating to cattle, by 348 to 79, indicating the decline in this crime after about 1900 and continuing in the interwar years.[52]

The difficulties of detecting poisoning in cattle were no less formidable than among humans, though, technically, by the 1870s and 1880s, arsenic was one of the easiest poisons to detect. Even where suspicions had been aroused, the carcass of a dead animal had often been picked over or partly devoured by dogs, jackals and vultures before it was recovered by the police. The surge in cattle poisoning occurred at a time when rinderpest was rife in many parts of India, and the symptoms of arsenic poisoning were hard to distinguish from those caused by cattle plague. It was even suggested that Chamars 'work[ed] under the cloak of [the] murrain', much as opportunistic arsenic poisoners supposedly used cholera epidemics as a cover for human murders.[53] Animal remains often reached the chemical examiner in such a state of decomposition that effective analysis was impossible. Like many vegetable poisons, *sui* left few lasting traces and was less easily detected than arsenic. The absence of any legal constraint on the sale of arsenic, which, according to the District Magistrate of Gorakhpur, was sold 'as openly in the market as rice or

---

[48] *ARCE (NWP), 1894,* 1A; *ARCE (NWP), 1900,* IA.
[49] *ARCED (Bengal), 1899,* 1, 3, 8.   [50] [Anon.], 'Poisoning', 642.
[51] *RCA (Bombay), 1891,* 2; *RCA (Bombay), 1908,* 2.
[52] *RCA (Bombay), 1921,* 1. The gulf continued to widen: in 1940, the Chemical Analyser investigated 47 animal cases to 781 human ones: *RCA (Bombay), 1940,* 1.
[53] 'Report on Cattle Poisoning', 651.

any other foodstuff', was a further obstacle to prevention and detection.[54] He argued that a 'stringent law regulating the sale of arsenic has been in force for many years in England without inconvenience to the public'. Why could not the same happen in India?[55]

There was a growing demand that there should be more effective control over the sale of arsenic and other poisons. Deterrent punishments should be imposed on cattle poisoners, and in all suspicious cases the stomach contents of cattle should routinely be sent to the chemical examiner for analysis.[56] By the late 1870s, the upsurge in cattle poisoning caused the government to believe that a poisons act might, after all, be necessary. The provincial administrations were consulted but they remained divided about the need for legislation, given the extensive commercial and medicinal uses of arsenic, the wide range of vegetable as well as mineral poisons available and the practical difficulties of implementing such a scheme in India. In 1881, the Government of India accordingly declined to act.[57] Three years later, in reply to L. A. Waddell, the Chemical Examiner, the Government of Bengal declared that legislation to regulate poisonous drugs was 'beset with many difficulties. Poisonous substances exist in every hedge and garden throughout the country, and in the present state of society in Bengal it seems quite impossible that the sale of poisons generally can be effectually controlled.'[58] Again the momentum for a poisons act stalled, and it was a further ten years before the issue was revived.

The Indian Poisons Act

In the mid-1890s, forty years after the idea of an arsenic act was first mooted, the Government of India was again forced to take up the issue. Several factors contributed to this revisiting. One was the unsuccessful prosecution in 1895 of Ellen Wagner and her lover, James Cray, for the murder of her husband, W. H. Wagner. His death at the Calcutta Medical College hospital was initially attributed to cholera but his exhumed body revealed arsenic as the most likely cause. Although the case was dismissed for want of sufficient evidence against the accused,

---

[54] Spedding to Commissioner, Benares Division, 11 October 1879, Bengal, Judicial, IOR.
[55] Ibid. On Spedding's role in reviving concern about cattle poisoning, see Mishra, 'Of Poisoners', 328–29.
[56] 'Report on Cattle Poisoning', 651.
[57] Extract from India, Home, Revenue, Agricultural (Judicial), no. 19, Bengal, Judicial, May 1881, IOR.
[58] Quoted in J. F. Evans and Chunilal Bose, 'The Necessity for an Act Restricting the Free Sale of Poisons in Bengal', in *Transactions of the First Indian Medical Congress, Calcutta, 24th to 29th December, 1894* (Calcutta: Caledonian Steam Printing Works, 1895), 478.

perhaps because this was a case involving Europeans and Eurasians, it revived official concern at the apparent ease with which arsenic could be obtained and used for homicidal purposes in India and highlighted the discrepancy between local laxity and the strict regulation of arsenic sales in Britain.[59] A further factor was the continuing flow of evidence from the courts, the police and the chemical examiners that arsenic was being used, on a relatively large scale, to commit murder and kill cattle.[60] But a third factor, perhaps the most significant of all, was the extent to which the Indian public and those Indians who spoke on their behalf, like Chunilal Bose as Bengal's Chemical Examiner, voiced alarm at the dangers posed by the careless and negligent sale of arsenic and other poisonous drugs in shops and bazaars. In effect, this was a restatement of the concern raised by Mouat as far back as 1843 – only now it was specifically targeted at arsenic, was supported by decades of accumulating statistics and reports and had the articulate backing of an emerging Indian (or at least Calcutta) public.

In the absence of an enquiry by the government itself, the single most important document presented in this renewed debate was a lengthy paper, 'The Necessity for an Act Regulating the Free Sale of Poisons in Bengal', jointly written by J. F. Evans of the IMS and Chunilal Bose, and presented to the first Indian Medical Congress when it met in Calcutta in December 1894.[61] Although the paper appeared under both their names, there is little doubt that Bose was the principal author. Indeed, the judge who presided at the Wagner-Cray trial and commended the Evans-Bose paper to the government, remarked that while he knew nothing of Evans, he believed Bose to be 'a very able man'.[62] Ostensibly confined to Bengal, the paper in fact gave a remarkably wide-ranging account of the use of arsenic, aconite, nux vomica and other 'deadly poisons' for the purposes of murder, suicide and cattle killing. At the same time, the familiar warning was given that, because of the difficulties of detection and the small number of cases referred to the chemical examiners, too little was known about the true extent of poisoning in India, which was assumed to be far greater than available statistics implied. Arsenic was central to this discussion. The importation of 30 or 40 tons of arsenic a year might not, the authors conceded, appear very considerable when compared to the 2,875 tons mined in Cornwall each year, but, given that as little as two

---

[59] John F. Norris to A. Miller, 9 May 1895, India, Home (Judicial), no. 80, August 1895, IOR. According to the *ToI*, 9 March 1895, 5, Ellen Wagner was an Indian Christian, her husband European, and Cray Eurasian.

[60] E.g., *ARCED (Bengal), 1900*, 6, 8–9.   [61] Evans and Bose, 'Necessity', 467–87.

[62] Norris to Miller, 9 May 1895, India, Home (Judicial), no. 80, August 1895, IOR.

grains of arsenic could kill a person, it was still a source of exceptional danger and legitimate concern.[63]

Two points made in Evans and Bose's paper were particularly compelling. One was that the data available for Bengal indicated that murder by poisoning was a far more common crime than in Britain, with three or four times the number of cases. In 1876–80, there had been twelve poison murders in England (equal to 0.07 per million of the population), but in Bengal for the same period ninety-four cases had been recorded (0.31 per million). Data for 1889–93 showed a slight decline (to eighty-one cases) but this still represented 0.23 per million.[64] The evidence with respect to suicide by poison was less easy to compute, but there, too, the authors believed that the incidence in Bengal was very high. In Calcutta, there had been 'a truly alarming increase in the crime of self-destruction', with one suicide death for every 1,300 of the population over the previous six years, and of these deaths 70 per cent were due to poisoning.[65]

The second telling point was with regard to accidental deaths by poisoning. Evans and Bose took the view that, given the predominantly rural nature of Bengali society, these were unlikely to be as high as in urbanized, industrialized Britain, where toxic substances of one kind or another were more readily available. In 1889–93, only eleven such deaths had been reported in Bengal, amounting to a mere 3.6 fatalities for every million of the population. But, even so, especially when non-fatal cases were added, the tally appeared to be large and growing. Annually, across Bengal, 395 fatal and non-fatal instances of accidental poisoning were reported: of these, all but fifty of the non-fatal cases resulted in recovery.[66] Some of these accidental poisonings occurred in the home or followed from 'secret preparations and quack remedies' sold to the 'poor and ignorant classes', including the 'love philtres' or aphrodisiacs 'in common use among the ignorant people of the country'. But blame fell, too, on the *bania*s and *pansari*s who sold dangerous drugs in the bazaars or who negligently failed to keep toxic substances like arsenic separate from the adjacent sacks of salt and sugar, with which they became mixed up or confused. These merchants and drug sellers were denounced by Evans and Bose for being 'as a rule, ignorant people, scarcely able to read or write their own language', and with a worryingly imperfect knowledge of what they sold to the public.[67] While many earlier poison panics, and more recent plague rumours and riots, had identified the British as a race of rulers intent on poisoning Indians and subverting their caste and religion, here was a clear statement of class fear and antipathy. It was

[63] Evans and Bose, 'Necessity', 481.   [64] Ibid., 468.   [65] Ibid., 469.   [66] Ibid., 470.
[67] Ibid., 477–78.

the 'poor and ignorant' purchasers of love drugs and the 'ignorant' and illiterate market traders who were to blame for endangering their own or other Indians' lives and for using, peddling or administering such lethal substances.

The Evans-Bose paper was in part a plea, as old as the appeals made by Mouat, Hervey and Chevers, for the more reliable collection and correlation of data relating to the supposedly widespread incidence of poisoning in India. It was an argument, too, for vesting still greater authority in the chemical examiners. But, citing the many lost opportunities and government prevarications since the 1840s, the authors also presented a powerful argument for legislative action along the lines of existing British laws and specifically for an Indian arsenic act. In making this case, however, they had to address the difficult issue of the medicinal use of arsenic by allopathic and indigenous medical practitioners – especially the latter at a time when the Ayurvedic revival was making indigenous medicine increasingly popular and prestigious. Evans and Bose tactfully sought to distinguish between 'ignorant' drug traders on the one hand and, on the other, the skilled and learned practitioners of indigenous medicine, who could not reasonably be denied the use of arsenic and other potentially dangerous substances, so long as these were duly labelled and sales suitably recorded.[68] Indeed, as a class, most hakims were said to be 'a respectable body of men', and it was not intended to interfere with their legitimate practice.[69] Considering how often *vaids* and hakims had been – and still were – reviled as 'quacks' and purveyors of toxic drugs, this was a singularly accommodating gesture, perhaps based on genuine conviction but also designed to forestall the opposition such an act would otherwise inevitably attract.[70]

Setting earlier objections aside, in August 1895, the Government of India again addressed provincial governments and commercial organizations across the country, this time inviting comments on an act specifically targeting arsenic and forwarding extracts from the Evans and Bose paper for consideration.[71] As Mark Brown has shown with respect to the Criminal Tribes Act of 1871, the deliberations of the Government of India could be long and tortuous and throw up strikingly different views on policy, before settling on an approved course of action – one that might then be embodied in law but still be subject to substantial modification over the years following its introduction.[72] This was similarly the case

[68] Ibid., 480–81.   [69] Ibid., 485.

[70] For continuing attribution of poisoning to hakims and 'quacks', see *ARCE (UP), 1937*, 6.

[71] J. P. Hewitt, Chief Secretary, India, to all governments, 26 August 1895, India, Home (Judicial), nos 82–83, August 1895, IOR.

[72] Mark Brown, *Penal Power and Colonial Rule* (Abingdon: Routledge, 2014).

with what, after nine years of consultation and debate, became the Indian Poisons Act of 1904. The arguments in favour of such an act did not greatly depart from those made by Evans and Bose in 1894 – the authoritative precedent of British legislation, the enormous (if inadequately reported) extent of poison crime in India, the dangers of accidental poisoning and the menace posed by the unregulated sale of arsenic. For many of those among the police, judiciary and medical profession who favoured action, the question of a poison act on British lines was intimately bound up with the need, also under consideration at this time, for a Medical Registration Act to regulate and license all practitioners. Bombay's Surgeon-General emphasized this point when he wrote that in India anyone could style himself a physician and practise medicine, 'no matter how ignorant he is', or sell dangerous drugs. 'At the end of the 19th century', he remarked, 'such a condition in a civilized country forming part of the British Empire, is an anomaly which only requires to be stated to make its removal a certainty'.[73] India, and not just arsenic, was on trial.

The case for legislative intervention met, however, with a barrage of opposition, and in this hostile and often ill-tempered response we can see foreshadowed some of the ways in which measures more generally to control poison (and latterly pollution) encountered resistance in India. Arguments against an arsenic act included the evident failure of the 1866 law in Bombay, the apparent unreliability of means for determining the criminal (as opposed to accidental) incidence of poisoning, the violation of the sacrosanct principle of free trade, the economic threat posed to the leather industry if the sale of arsenic were restricted, the problem of establishing a reliable inspection agency that would not itself become an engine of oppression and corruption, the ignorance of bazaar traders and the impossibility of outlawing the criminal use of arsenic without interfering with its legitimate use. Some of the most strident voices raised in opposition came from Calcutta. One representative of the city's commercial interests wrote: 'I do not think it advisable to burden our statute-book with complicated laws tending to vex people, increase their responsibilities, and interfere with their actions, because a few of them use their liberty to the prejudice of others.' If poisons were to be banned, he continued, why not the knives, sticks, 'and the thousand and one instruments used by criminals for their felonious purposes'?[74] The Secretary of the Calcutta Trades Association took a more material stance. Restricting the sale of arsenic, he wrote, would hit those who used it medicinally, not

---

[73] Henry Cooke to Secretary, Bombay, Judicial, 16 March 1896, India, Home (Judicial), nos 428–42, May 1899, IOR.
[74] R. Braunfeld to Secretary, Calcutta Corporation, 2 December 1895, ibid.

just doctors but also the railway, colliery, jute mill and plantation managers who bought Fowler's Solution (which contained arsenic) to protect their workers against malaria. If other potential poisons, such as corrosive sublimate and 'strong acids' like nitric and sulphuric acid, were also brought under a poisons act, photographic studios would suffer, the manufacturers of electric batteries would be penalized and soda water factories would go out of business.[75] While no one actually defended poisoning, it was already becoming clear in the 1890s and 1900s that business and industrial interests were adamantly opposed to measures which they saw as restricting their commercial freedom and economic profitability.

There were dissenting voices, too, from within the colonial bureaucracy. E. H. C. Walsh, Secretary to the Board of Revenue in Bengal, was scathing about the proposed bill. He was unimpressed by Evans and Bose's paper 'as it is so obviously written by men either ignorant of, or ignoring, the conditions of Indian life'. They had greatly exaggerated the dangers of poisoning, and the 'impracticality' of their proposals was, Walsh believed, revealed when they tried to detail how such an act would function. He went on: 'There is no doubt that one of the banes of English rule in India is the proneness to listen to well-intentioned faddists, without sufficiently taking into consideration the effect of complying with their views.' No one would deny that it was a 'good thing' for poisons to be kept out of the hands of 'would-be murderers and suicides', but it was possible in carrying out such a theory 'to do more harm than good'.[76]

Given such hostility, it is perhaps surprising that the Government of India persisted in its course of action and in 1903 drafted an act for the Secretary of State's approval. In the final analysis, the government was swayed more by the continuing threat to human life and the loss of cattle than by objections from the leather trade and other commercial interests. Perhaps, the District Magistrate of Burdaun captured the underlying concern when he remarked in 1896 that there was 'a general horror of arsenic'.[77] In giving his approval to the act, the Secretary of State, Lord George Francis Hamilton, seemed swayed by two considerations. One was the continuing confusion between cholera deaths and arsenic poisoning which implied stricter control over poison sales so that such crimes could be more adequately detected and punished.[78] The second was that

[75] G. Hickie to Secretary, Bengal, Municipal (Medical), 25 February 1896, ibid.
[76] E. H. C. Walsh to Secretary, Bengal, Municipal (Medical), 4 February 1896, ibid.
[77] G. A. Tweedy to Commissioner, Rohilkhand, 10 January 1896, ibid., nos 570–75.
[78] There were many other examples of supposed cholera deaths that proved, on exhumation, to be due to arsenic: for example, *RCA (Bombay), 1875–76*, 18.

this was an issue that touched upon the standing of India as a civilized country and on the moral legitimacy of empire. There was, he averred, 'no other civilized country in the world in which the sale and possession of poisons is not carefully restricted'.[79] In the statement of objects and reasons presented by the Government of India in July 1903, prominence was accordingly given to the 'extraordinary anomaly' of India being almost alone in the civilized world in not having legislation to control 'a deadly poison, such as arsenic', which remained 'procurable in unlimited quantities in every part of [the country]'.[80]

The toxic watershed

The Indian Poisons Act of 1904 contained much the same provisions as the Bombay act of 1866 and the British pharmacy acts which had preceded it.[81] Sellers of poison were required to enter in a register every sale made and to note the type and quantity of poison sold, the name and address of the purchaser and the intended use of the poison. Both vendor and purchaser were to endorse the entry with their signature or thumbprint. Poison was to be kept in a box, prominently marked 'poison' in red in English and the vernacular and displaying a warning skull and crossbones. The substances notified under the schedule included white arsenic and the various sulphides and compounds of arsenic as well as aconite, strychnine, nux vomica and corrosive sublimate.[82]

Within a few years, however, the 1904 act, like the Bombay law before it, was being described as 'tentative and defective', with many druggists avoiding registration or remaining in ignorance of the act's provisions.[83] It required a further amendment in 1919 to strengthen the act and restrict access to arsenic and other poisons by medical practitioners.[84] The number of recorded offences and convictions under the act

---

[79] Lord George Francis Hamilton to Governor-General, 10 April 1903, India, Legislative, no. 16, January 1904, IOR.

[80] Appendix L, 'Statement of Objects and Reasons', ibid.

[81] Peter Bartrip, ' "A Pennurth of Arsenic for Rat Poison": The Arsenic Act 1851 and the Prevention of Secret Poisoning', *Medical History* 36 (1992): 53–69; Ian Burney, *Poison, Detection, and the Victorian Imagination* (Manchester: Manchester University Press, 2006), 64–65.

[82] Bengal, Municipal (Medical), no. 25, September 1909, IOR.

[83] H. L. Stephenson, Secretary, Bengal, Financial, to Secretary, India, Home, 20 April 1915, India, Home (Judicial), no. 167, May 1918, IOR; Bengal, Municipal (Medical), nos 59–63, May 1909, IOR.

[84] India, Home (Judicial), nos 20–34, March 1907, IOR; Bengal, Municipal (Medical), nos 23–33, April 1910, IOR; India, Home (Judicial), nos. 118–19, January 1911, IOR; India, Home (Judicial), nos 253–54, February 1920, IOR. For the amended act, see *Poisons Manual, 1934* (Calcutta: Superintendent of Government Printing, Bengal, 1934).

remained small – only twelve cases were brought before Bengal's criminal courts in 1914 (none of which led to a conviction) and eight in 1920 (of which seven resulted in convictions).[85] Even so, the conditions imposed by the act appear to have discouraged arsenic sales and to have had a marked effect on the number of accidental deaths, homicidal poisonings and cattle killings attributed to arsenic.[86] By the 1930s, poisoning by means of arsenic had become far less common, while, according to the chemical examiners, there had been a corresponding increase in the criminal use of opium and datura, with poison cases involving these poisons now two or three times more frequent than those involving arsenic. The number of arsenic cases seen by Bombay's Chemical Analyser dropped from fourteen in 1931 to six in 1945, while opium investigations only fell from eighty-eight to seventy-four and datura from twenty-two to twelve over the same period.[87] According to UP's Chemical Examiner, one of the main reasons for this shift away was criminals' 'increasing knowledge of the fact, with the increased spread of education, that the organic poisons give rise to less characteristic symptoms [than inorganic substances] and are less readily detected afterwards'.[88]

And there a history of poisons and poisoning in British India might end – with the 1904 Poisons Act, its subsequent amendment and the decline of arsenic poisoning in India. Such an ending might be taken to imply that colonialism had eventually, after many decades, managed to resolve an issue of criminal toxicity which, materially and conceptually, it had helped bring into being in the first place. Or that history might conclude with Ram Nath Chopra, whose patriotic investigation into India's narcotic drugs, poisonous plants and venomous snakes between the 1930s and 1960s complemented and extended the pharmacological and toxicological enquiries of earlier British and Indian researchers from W. B. O'Shaughnessy to Chunilal Bose.[89] Chopra might be seen as helping to bring about a nationalist resolution of the long-standing question of Indian poisons, their use, misuse and appropriate governance.

---

[85] *Report on the Administration of Criminal Justice in the Presidency of Bengal, 1914*, 21; *Report on the Administration of Criminal Justice in the Presidency of Bengal, 1920*, 21.

[86] This decline parallels that in England and Wales from the 1850s onwards: Watson, *Poisoned Lives*, 42–43.

[87] *RCA (Bombay), 1931*, 2; *RCA (Bombay), 1945*, 3.   [88] *ARCE (UP), 1937*, 6.

[89] R. N. Chopra, *Indigenous Drugs of India: Their Medical and Economic Aspects* (Calcutta: N, Mukherjee, 1933); R. N. Chopra, R. L. Bahadur and S. Ghosh, *Poisonous Plants of India* (2 vols, New Delhi: Indian Council of Agricultural Research, 1965). On Chopra, see Pratik Chakrabarti, *Bacteriology in British India: Laboratory Medicine and the Tropics* (Rochester, NY: Rochester University Press, 2012), 136–40.

But there are strong grounds for arguing that neither the passing of the 1904 act nor the interwar period of scientific investigation represented closure so much as the waning of one phase in the history of Indian toxicity and the dawning of another. Poison's penumbra was widening, not contracting, and, as the remaining part of this book will try to show, the problem of poisons and poisoning, far from disappearing, took on new meanings and contexts. The fear of poisoning and the scares surrounding its real or imagined presence did not abruptly cease, but became subsumed into a new language of toxicity, adulteration, contamination and pollution, and entered a new era of public, state and scientific anxiety. The specialist agencies which had been in part created to detect homicidal poisoning – the chemical examiners among them – did not suddenly forfeit their significance but instead took on an expanding role and increasing workload as the parameters of toxicity widened still further.[90] While homicide, suicide and other forms of calculated poisoning might no longer stand quite so high on the scientific agenda, many of the same substances that perturbed nineteenth-century toxicology – arsenic, mercury and lead – resurfaced in a new era of human poisoning and environmental pollution in which individual culpability and criminal intent were less easily established. For all its limitations, the 1904 act occasioned – or, more accurately, coincided with – a new sense of engagement with, and responsibility for, toxicity in India. Poison cast an ever-lengthening shadow.

In this transition from one poison register to another, Chunilal Bose, as the Chemical Examiner for Bengal between 1889 and 1920, was a seminal figure. This was not just for co-authoring with Evans the paper on which the 1904 act was eventually based: it was also for his assumption of public leadership on several poison- or pollution-related matters. No single episode better illustrates this role than the Bhowanipore food poisoning incident. Bhowanipore was a prosperous suburb of Calcutta, lying to the south-east of the central maidan. A feast was held in the grounds of a large house there over three days in late June 1903, attended by the 'cream of ... Bengali Society', men and women from the leading *bhadralok* families of the city.[91] On 27 June, the second evening, more than 600 guests were present, dining in batches between 8.30 and 11 pm.

---

[90] Thus, the number of poison cases reviewed by UP's Chemical Examiner increased tenfold between 1871–80 and 1931–40 (from an average of 342 a year to 3,584), with the greatest increase from 1921–30 to 1931–40: *ARCE (UP), 1938*, 2.

[91] Chunilal Bose, 'The Bhowanipore Food-Poisoning Case', in J. P. Bose (ed.), *The Scientific and Other Papers of Rai Chunilal Bose Bahadur* (2 vols, Calcutta: Forward Press, 1924), 1: 301. *Bhadralok* refers to Bengal's Hindu elite drawn from the Brahmin, Vaidya and Kayastha castes.

About a quarter of those present that evening rapidly developed symptoms of acute gastroenteritis: dozens remained ill for days and four died.
The guests who dined on the other days, 26 and 28 June, were
unaffected.[92] Four months later, a short article on the 'Bhowanipore
food poisoning case' appeared in the *Indian Medical Gazette* written by
J. Neild Cook, the city's Health Officer. Cook noted that food poisoning
had seldom been investigated and 'worked out bacteriologically' in India,
but thought it was probably more common there than in 'more temperate
climates'.[93] He considered possible causes of the outbreak, observing that
some form of colon bacillus, perhaps derived from the tank water in which
the serving dishes and pots had been washed, was responsible. But he
then added that this 'unfortunate occurrence' might not be an 'unmixed
evil' if it 'brings home to the educated native of this country that hygiene
has a practical as well as a theoretical side, as he is very ready to talk about
sanitation but too frequently fails to apply the principles he preaches to
the everyday affairs of life'.[94]

Early in 1904, Chunilal Bose responded fiercely with a 35-page article
in the *Calcutta Practitioner*.[95] His fury had several causes. One was that the
food poisoning episode had affected him personally. Unlike Cook, he had
been present as a guest at the feast on 27 June and been ill for several days
as a result. More than that, a nephew of his, also present that evening,
subsequently died from food poisoning. Second, Bose had collected the
testimony of thirty-six of the individuals who had been affected, and he
reconstructed from their evidence (and his own experience) the source of
the poisoning. Pitting his scientific expertise against that of Calcutta's
Health Officer, Bose established that it was the ice cream that was the
culprit. Third, Cook had earlier offended middle-class sensibilities in
Calcutta by his enforcement of unpopular anti-plague measures, and
the concluding remarks in his article clearly irked Bose still further. At a
time when Bose was campaigning for sanitary reform in Bengal, Cook's
suggestion that Indians treated hygiene as a 'theoretical' rather than
'practical' matter must have felt like a direct snub to his own scientific
credentials and social leadership.[96] Certainly, it was an insult to the
Bengali *bhadralok*, which, as an enlightened and progressive body of
men and women, surely did not need lessons from Cook on hygiene

---

[92] Bose, 'Bhowanipore', 301.
[93] J. Neild Cook, 'The Bhowanipore Food-Poisoning', *IMG* 38 (1903): 362.
[94] Ibid., 364.   [95] Bose, 'Bhowanipore', 298–333.
[96] Bose's campaigning led, after 1918, to a plea for India's 'sanitary reconstruction':
Chunilal Bose, 'A Few Hints on Sanitary Reconstruction', in Bose, *Scientific Papers*, 2:
175–91.

and sanitation.[97] It remained, however, to explain how the ice cream had become infected. 'The fact of Dr Cook's finding some indefinite colon bacilli in one of the foulest tanks in Bhowanipore does not help', Bose wrote, 'to establish its connection with the infection of the ice-cream. We must look somewhere else for the source of the infection.'[98] In exculpating the Bengali *bhadralok* from blame, Bose pointed an accusing finger instead at the low-caste stable boys and grooms who had helped prepare the fatal ice cream. The gastroenteric 'toxin' might, he speculated, have originated in the soil or in horse dung and been communicated to the milk, and the ice cream made from it, by the grooms working the ice cream machine. Grooms, Bose added, 'are not reputed to be very cleanly in their habits', and from their hands 'a weighable quantity of horse-dung can always be scraped out'.[99] The need to combat the adulteration of Calcutta's milk supply with polluted water (containing typhoid and other dangerous 'germs') was one of the many public health causes Bose had espoused, describing it as a cause of 'terrible mortality' among Bengali women and children in particular.[100] Like the food poisoning at Bhowanipore, milk adulteration was one of the ways in which the health and security of the *bhadralok* home was seen to be under threat – much as Europeans feared that Indian servants brought poison and disease into their homes. The effect was to deflect responsibility for poisoning away from the 'respectable' classes (European or Indian) and onto the low-caste servants, milkmen and grooms who worked for them, handled their food and tainted their milk. Poisoning was increasingly taking on the contours of class.

Bhowanipore exemplified a wider process of change. Thereafter, food poisoning – whether caused by bacteria or as a result of accidental contamination with rat poison, pesticides and other toxic substances – attracted growing attention in the reports of the chemical examiners and municipal bacteriologists, in the Indian medical literature and in the local press. Although food of all kinds came under increased scrutiny, the widespread adulteration and contamination of milk and milk products was a primary concern. Bovine tuberculosis was rare, but typhoid, cholera and other bacillary or parasitic infections were transmitted through the

[97] On the spread of sanitary ideas among the Bengali elite, see Pradip Kumar Bose (ed.), *Health and Society in Bengal: A Selection from Late 19th-Century Bengali Periodicals* (New Delhi: Sage, 2006), ch. 4.

[98] Bose, 'Bhowanipore', 330.   [99] Ibid., 333.

[100] Chunilal Bose, 'The Milk-Supply of Calcutta: Its Hygienic, Commercial and Social Aspects', in Bose, *Scientific Papers*, 2: 113–60. On this issue, see Mishra, *Beastly Encounters of the Raj*, ch. 5. That milk adulteration was not an issue unique to India is evident from P. J. Atkins, 'White Poison? The Social Consequences of Milk Consumption, 1850–1914', *Social History of Medicine* 5 (1992): 207–27.

polluted water used to dilute milk, and in Bombay, where milk adulteration was rife, bacteriological analysis routinely revealed evidence of dilution or showed a 'very high degree of manurial and other undesirable
pollution'.[101] In 1905, Bombay's municipal laboratories examined 1,701
items: among these were 223 samples of milk (of which at least 130 had
been subject to dilution) as well as adulterated butter and ghee.[102] By the
late 1940s, 'food poisoning' was listed in the chemical examiners' reports,
along with arsenic, opium and datura, among the principal causes of
poison cases and poison-related deaths.[103]

There were other ways, too, in which the idea of poisoning was undergoing subtle redefinition by the 1930s and 1940s. Thus, for most of the
colonial era, there was a reluctance to regard alcohol as a major source of
poisoning, though adding datura and other drugs to arrack to increase its
potency was sometimes considered a danger, especially to British
soldiers.[104] However, in the late 1930s, as prohibition became official
policy in provinces under Congress Party rule, so alcohol poisoning began
to assume unprecedented prominence in the data presented in the chemical examiners' reports. This innovation perhaps reflected a new clinical
awareness or a change in police practice towards now illicit drinking.
Conceivably, it underscored the idea that drinking was now, under a
nationalist administration, to be viewed as socially undesirable and
morally reprehensible, a legacy of a decadent, drink-indulgent colonialism. More likely, though, it resulted from the way in which the official ban
on alcohol, aggravated by poverty, drove determined drinkers to consume
hooch containing a toxic mix of improvised ingredients from cleaning
fluid and varnish to industrial alcohol. The trend became even more
marked after independence as the policy of prohibition was implemented
more widely. Of the 4,595 suspected poison cases referred to Bombay's
Chemical Analyser in 1951, 1,900 related to alcohol poisoning.[105] Here
was the start of a toxic trend with which post-independence India has
been sadly all too familiar.[106]

---

[101] RHO, *ARMC (Bombay), 1904–05*, 187–89; RHO, *ARMC (Bombay), 1914–15*, 125.
Laboratory analysis showed that in Calcutta the quality of the milk supply had not
greatly improved by the 1940s: R. Banerjea and A. K. Sen, 'The Bacterial Content of the
Calcutta Milk Supply', *IMG* 81 (1946): 40–45.
[102] RHO, *ARMC (Bombay), 1905–06*, 195.   [103] *RCA (Bombay), 1949*, 2.
[104] As in early-nineteenth-century Bengal: F/4/611: 15227, IOR.
[105] *RCA (Bombay), 1951*, 3.
[106] In Bangalore in July 1981, more than 300 people died from alcohol poisoning. Manor
points out that 2,000 such deaths were reported between 1972 and 1992, implying there
were many others that escaped press attention: James Manor, *Power, Poverty and Poison:
Disaster and Response in an Indian City* (New Delhi: Sage, 1993), 18.

Tropical poisons

Chunilal Bose's reference to 'toxins' in relation to the Bhowanipore food poisoning episode should alert us, too, to the profound transformation taking place in medicine and public health in India in the 1890s and 1900s, and to further changes in the scientific and public understanding of what constituted a poison. In order to explain this, it is first necessary to backtrack a little. In the eighteenth and nineteenth centuries, under the influence of neo-Hippocratic thought and the revived emphasis on 'airs, waters and places', it had been common for medical writers in India, as in Europe, to explain disease in terms of either ailments whose action appeared to mimic that of poison or which were seen to result from some actual poison entering the body. One perceived source of these poisons was the miasma that resulted from the decay of vegetable and animal matter or emanated from dense vegetation, stagnant water and crowded human habitations, accentuated (as in India) by heat and humidity.[107] In discussing fevers in 1828, James Annesley in Madras dwelt on the danger of 'insalubrious exhalations' from swamps, rice fields, river banks and 'low jungles'. 'The sum of our knowledge ... of the poison', he wrote, 'seems to be that it proceeds from those elements which exist in a rich soil and nourish the vegetable and animal kingdoms'.[108] While decomposition might anywhere generate poisonous miasma, the effect was perceived to be greater in tropical regions where putrefaction was more rapid and intense. Moist, hot tropical air, Annesley explained, 'holds in solution that principle or effluvium which is formed from the soil, and which, contaminating the air, produces endemic and epidemic diseases'.[109] A decade later, also discussing malarial fevers, F. P. Strong in Bengal similarly observed that jungles, lakes, marshes and stagnant pools of water gave rise to a 'constant supply of the poison'. The chemical composition of that 'poison' had yet to be determined, but it was clear that it was 'produced and propagated by wet soils, hot countries, a particular state of atmosphere, and high temperature'.[110]

Miasmatic theory proved remarkably stubborn. As late as 1886, a former surgeon-general, W. C. Maclean, described malaria as 'an earth-born poison, for the most part generated in soils the energies of which are

---

[107] Harish Naraindas, 'Poisons, Putrescence and the Weather: A Genealogy of the Advent of Tropical Medicine', *Contributions to Indian Sociology* 30 (1996): 1–35.

[108] James Annesley, *Researches into the Causes, Nature, and Treatment of the Most Prevalent Diseases of India* (2 vols, London: Longman, Rees, Orme, Brown & Green, 1828), 1: 75.

[109] Ibid., 1: 5–7, 70–71.

[110] F. P. Strong, *Extracts from the Topography and Vital Statistics of Calcutta* (Calcutta: no publisher, 1837), 10.

not expended in the growth and sustenance of healthy vegetation'.[111] But between the mid-1880s and early 1900s, the rise of bacteriology and the germ theory (or theories) of disease had begun to demonstrate in India, as elsewhere, that, rather than some elusive poison, specific microorganisms were responsible for causing diseases like cholera, tuberculosis and typhoid. Bacteriology provided the means of detecting the identity of pathogenic microorganisms that had previously been categorized (for want of a better understanding) as 'poisons'. Parasitology and protozoology further demonstrated that the malarial 'poison' to which Annesley and Strong alluded was actually a protozoan parasite transmitted by *Anopheles* mosquitoes.

Changing concepts of toxicity and new ideas about how poisons might be analysed and used were central to the way in which this revolution in medical science impinged upon India. For instance, in the wake of Louis Pasteur and Robert Koch, practitioners of the new bacteriology were able to demonstrate how 'poisons' like the cholera bacillus could be modified in the laboratory to produce prophylactic or therapeutic vaccines. This principle of turning destructive poisons into curative potions was given practical effect in India in the 1890s and 1900s by the Russian-born bacteriologist Waldemar Haffkine, who developed vaccines for use against cholera and plague.[112] Somewhat earlier, in 1884, two IMS officers based at the Calcutta Medical College, C. J. H. Warden and L. A. Waddell, took advantage of the visit to India of Koch and the German Cholera Commission to investigate the properties of the *Abrus* seeds used in cattle poisoning. It had recently been claimed by French researchers that the 'toxic effect' of the seeds was caused by a bacterial infection. Warden and Waddell used techniques of bacteriological and chemical analysis to show conclusively that the presence of bacteria at the site of wounds resulting from the *sui* spike was merely incidental. The 'toxic action' was due to a previously unknown 'proteid', which they called abrin, and whose physiological effects, like those of snake venom, only became manifest when injected subcutaneously and it entered the human or animal bloodstream.[113]

The bacteriological revolution helped establish and propagate a heightened language of toxicity. Instead of miasmatic poisons, bacteriologists

---

[111] William Campbell Maclean, *Diseases of Tropical Climates* (London: Macmillan, 1886), 25.

[112] Chakrabarti, *Bacteriology*, 11, 41; J. D. Graham, 'Medical and Research Organisation', in Far Eastern Association of Tropical Medicine, *The Indian Empire* (Calcutta: Thacker, 1927), 99–103.

[113] C. J. H. Warden and L. A. Waddell, *The Non-Baciliar Nature of* Abrus-*Poison, with Observations on Its Chemical and Physiological Properties* (Calcutta: Bengal Secretariat Press, 1884).

now spoke of 'toxins' – as did Chunilal Bose in relation to the food poisoning at Bhowanipore. In 1913, Patrick Hehir, havering somewhat between old and new, explained that milk adulterated with polluted water could cause severe cramps and diarrhoea due to 'a special chemical poison of non-bacterial origin called *lacto-toxin* or *tyro-toxin*'.[114] By 1914, it had become common for what had previously been described as 'poisons' – like the causative mechanism behind cholera or snakebite – to be rebranded as 'toxins' and for their effects on the human body to be re-conceived as 'intoxication'.[115]

A further expansion of the toxic principle – in both its positive and negative connotations – came from the concurrent rise of chemotherapy. In order to combat disease within the body, this technique used various chemical substances, including those that had long been treated as having poisonous, as well as therapeutic, properties, like arsenic, mercury and antimony. One of the earliest instances of this – Paul Ehrlich's discovery of Salvarsan, an arsenic compound, in 1909 – provided the first medically effective treatment for syphilis. From an Indian perspective, there was a certain irony in this, since the use of arsenic by indigenous medicinal preparations had long been criticized as dangerous and unproductive by Western physicians and pharmacologists. The introduction of Salvarsan in 1913 helped reduce levels of venereal disease among British troops stationed in India, large numbers of whom had previously been hospitalized for treatment.[116] But the converse of this was a growing volume of reports of the toxic effects of intravenous injections of Salvarsan (and less toxic replacements like Neosalvarsan) due to what was, in practice, acute arsenicosis or arsenic poisoning, occasionally resulting in the death of patients undergoing treated for syphilis.[117] Barely had the nineteenth-century phenomenon of homicidal arsenic poisoning subsided than arsenic poisoning resurfaced in twentieth-century chemotherapy.[118]

---

[114] Patrick Hehir, *Hygiene and Diseases of India: A Popular Handbook* (3rd ed., Madras: Higginbothams, 1913), 297.

[115] D. D. Cunningham, 'The Physiological Action of Cobra-Venom', in *Scientific Memoirs by Medical Officers of the Army of India, Part XI* (Calcutta: Office of the Superintendent of Government Printing India, 1898); Hugh W. Acton and R. N. Chopra, 'The Nature and Pharmacological Action of Cholera Toxin', *IJMR* 12 (1924–25): 235–50.

[116] *Annual Report of the Sanitary Commissioner with the Government of India, 1913* (Calcutta: Superintendent, Government Press, 1915), 22.

[117] K. G. Pandalai, 'A Case of Salvarsan Poisoning', *IMG* 49 (1914): 59; J. P. Cullen, 'A Fatal Case of Poisoning by Neo-Karsivan', *IMG* 59 (1924): 245; A. Denham White and Sital Chandra Dutt, 'A Note on the Toxic Symptoms of Organic Arsenic', *IMG* 60 (1925): 464–65.

[118] Jaising P. Modi, *A Text-Book of Medical Jurisprudence and Toxicology* (2nd ed., Calcutta: Butterworth, 1922), 445–46. An advertisement for 'Sulfarsenol' in 1933 promised that it was 'the safest arsenical' and 'practically eliminated' the risk of arsenic poisoning: *IMG* 68 (1933): lxiii.

The problem was not resolved until the introduction, decades later, of antibiotics. In the meanwhile, toxicity became part of the rubric by which physicians and clinical researchers evaluated the effects of medicinal drugs and their undesirable, but seemingly unavoidable, consequences. It now became a question not of whether poisonous substances could be jettisoned entirely, but of how it might be possible to deliver a chemical dose to the patient that ensured maximum toxicity for the parasite while ensuring minimal toxicity to the human host.[119]

Toxicity began to figure more diversely in the investigation of India's tropical diseases and clinical responses to them. Indeed, there was a singular connection between the perceived toxicity of tropical diseases and the bacteria, parasites and insect vectors that harboured and transmitted them and the toxic ingredients needed to bring about their treatment, prevention and cure. One of the most troublesome tropical diseases as far as eastern India was concerned was kala-azar.[120] A form of leishmaniasis, kala-azar spread rapidly in Bengal and Assam from the 1850s onwards, causing widespread suffering and heavy mortality. Medical understanding of kala-azar was complicated by confusion with malaria and beriberi before it was established in 1903 that the disease was caused by a parasite spread by sandflies. This discovery began a frantic quest for an effective treatment. Various chemicals were tried, including arsenic.[121] However, the greatest success was obtained through intravenous injections with antimonial drugs. This technique was pioneered in India by Leonard Rogers, IMS, using tartar emetic, which contained antimony, and further developed from 1920 onwards by Upendranath Brahmachari at the University of Calcutta using a safer combination of antimonial salts.[122] Here was another illustration of how a poisonous substance (as antimony had long been regarded) could be used against tropical disease.[123] Like the use of arsenic against syphilis, however, antimonial

[119] J. F. Caius and K. S. Mhaskar, 'The Correlation between the Chemical Composition of Anthelmintics and Their Therapeutic Values in Connection with the Hookworm Inquiry in the Madras Presidency', *IJMR* 7 (1919–20): 430.

[120] Achintya Kumar Dutta, 'Medical Research and Control of Disease: Kala-Azar in British India', in Biswamoy Pati and Mark Harrison (eds), *Society, Medicine, and Politics: Colonial India, 1850–1940s* (London: Routledge, 2009), 93–112.

[121] F. Roux, 'Arsenic in the Treatment of Kala-Azar', *IMG* 48 (1913): 132–33; Sarasi Lal Sarkar, 'The Action of Quinine and Arsenical Preparations in Kala-Azar', *IMG* 50 (1915): 92–94.

[122] Upendranath Brahmachari, 'Campaign against Kala-azar in India', in Upendranath Brahmachari, *Gleanings from My Researches* (2 vols, Calcutta: University of Calcutta, 1940), 1: 420–27.

[123] Theodric Romeyn Beck and John R. Beck, *Elements of Medical Jurisprudence* (5th ed., London: Longman, Rees, Orme, Brown, Green & Longman, 1836), 786–90; Editorial, 'Poisoning by Antimony', *IMG* 38 (1903): 383–84. On the toxic side effects of

drugs could produce devastating, even fatal, side effects in patients. Antimony poisoning, previously rare in India, now became more common, and the question again had to be posed: how much suffering was acceptable in order to achieve a cure?[124] But Brahmachari's work on the antimonial treatment of kala-azar signalled more than a therapeutic breakthrough: it further established the international credentials of Indian science and of Indians in science. Brahmachari himself hailed his discovery as 'one of the most remarkable feats in chemotherapy'.[125]

Beriberi was another tropical disease enmeshed in new thinking about toxicity. During the nineteenth century, it had been periodically reported in India without becoming an object of sustained enquiry. International interest in the disease was sparked by its greater prevalence in South-East Asia, chiefly among migrant labourers, prisoners and asylum inmates. In the 1880s and 1890s, beriberi was still conceived in broadly miasmatic terms. It was designated a 'poison', and even characterized as 'a malarial poison arising from the decomposition of organic matter in the soil, favoured and strengthened by damp and moisture, and ... inhaled into the system through the lungs'.[126] An alternative, if short-lived, theory was that it was caused by arsenic poisoning since minute traces of that metalloid had been found in the hair of beriberi victims in Assam.[127] However, by 1910, research by Dutch, British, French and American researchers in South-East Asia had homed in on the nutritional deficiency resulting from a diet consisting almost exclusively of polished white rice. It is now known that beriberi is caused by removal of the grain's outer layers, and hence of its vitamin B1 content, due to mechanized milling and polishing.[128] But there remained epidemiologists in India who believed that this was an inadequate explanation, maintaining instead that it was caused by a 'specific poison', a 'volatile toxin' that originated outside the body, perhaps in the soil or in human habitations, and was akin in its action to poisoning by arsenic, lead or alcohol.[129] J. W. D. Megaw of the IMS argued for nearly twenty years that beriberi was due to a 'toxi-infection' either present in the grain or occurring as an ergot when rice

antimonials in kala-azar, see T. C. Boyd, L. Everard Napier and A. C. Roy, 'The Distribution of Antimony in the Body Organs', *IJMR* 19 (1931): 285–94.

[124] A. N. Bose, T. N. Ghosh, S. N. Mitra and S. Dutta, 'On the Toxicity of Some Organic Antimonial Drugs Used for the Treatment of Kala-Azar', *IMG* 81 (1946): 13–16.

[125] Brahmachari, 'The Conquest of Kala-Azar', in *Gleanings* 1: 461.

[126] T. Irvine Powell in Arthur J. M. Bentley, *Beri-Beri: Its Etiology, Symptoms, Treatment, and Pathology* (Edinburgh: Young J. Pentland, 1893), 13.

[127] Editorial, 'Dr H. Durham's Report on Beriberi', *IMG* 39 (1904): 221–22.

[128] K. Codell Carter, 'The Germ Theory, Beriberi, and the Deficiency Theory of Disease', *Medical History* 21 (1977): 119–36.

[129] Hehir, *Hygiene*, 591–93.

was stored for any length of time in damp conditions. This view ran counter to the growing international consensus that beriberi was a nutrition deficiency disease. But it encapsulated the idea, still powerful in many European minds, that toxicity in some form or other was omnipresent in the Indian environment, in Indian diets or in the manner in which food was stored, prepared and eaten.[130] It was not until the 1930s, after Megaw had left the country, that the 'intoxication' theory of beriberi was finally scotched and the nutritional explanation fully vindicated.[131]

An additional complication in the investigation of beriberi in India was its confusion with epidemic dropsy. The latter was a disease that medical commentators like Chevers in the 1880s and Megaw in the 1920s believed was 'closely related to, if not identical with', beriberi.[132] In eastern India, where it was most prevalent, the disease, which caused swelling of the lower limbs, sparked a partial return to earlier poison panics, for it appeared to be related to the consumption of certain foodstuffs, especially rice.[133] In the Calcutta outbreak of 1909–10, where there were 1,581 cases and 368 deaths, it was observed that epidemic dropsy seemed to target high-caste Bengalis, but to have no corresponding effect even on other communities like the Marwaris.[134] Commenting on this apparent discrimination, two Bengali writers remarked that 'out of the many races of rice eaters this woeful malady has particularly picked up the Bengalee race as its most suitable victim'.[135] One explanation for epidemic dropsy was to relate it back to the 'toxi-infection' theory of rice and to regard it, like beriberi, as a 'poison ... formed during the seasoning process'.[136] In other words, beriberi and epidemic dropsy were manifestations of the same 'toxic syndrome caused by the ingestion of poisonous bases formed in rice under certain conditions'.[137] It was only in

[130] J. W. D. Megaw, 'The Beriberi Problem', address to the Far Eastern Association of Tropical Medicine, September 1923, in J. W. D. Megaw, *Collected Papers*, Wellcome Library, London.

[131] David Arnold, 'British India and the Beriberi Problem, 1798–1942', *Medical History* 54 (2010): 295–314.

[132] Norman Chevers, *A Commentary on the Diseases of India* (London: J. & A. Churchill, 1886), 60–79; J. W. D. Megaw, 'Notes on Cases of the "Epidemic Dropsy" Type of Beri-Beri at the Presidency General Hospital, Calcutta', *IMG* 45 (1910): 8.

[133] For the panic caused by the 1909–10 Calcutta outbreak, see Sarat Chandra Ghose, *Beri-Beri: Its Causation, Prevention and Homeopathic Treatment* (Calcutta: A. C. Dutt, 1910), 30.

[134] E. D. W. Greig, *Epidemic Dropsy in Calcutta (Final Report)* (Calcutta: Superintendent of Government Printing, India, 1912).

[135] S. C. Seal and M. N. De, 'Epidemic Dropsy', *Indian Medical Record* 58 (1937): 120–28.

[136] Hugh W. Acton, 'The Causation of Epidemic Dropsy', *IMG* 57 (1922): 333.

[137] Hugh W. Acton and R. N. Chopra, 'The Problem of Epidemic Dropsy and Beriberi', *IMG* 60 (1925): 17; J. W. D. Megaw, S. P. Bhattacharji and B. K. Paul, 'Further Observations on the Epidemic Dropsy Form of Beri-Beri', *IMG* 63 (1928): 418.

the 1930s that epidemic dropsy was conclusively distinguished from beriberi, when its cause was shown to be a very different kind of poisoning from that envisaged by earlier epidemiologists. Epidemic dropsy was attributed to the contamination of mustard oil, widely used in Bengali cooking, by argemone oil, which contained a toxic alkaloid. This occurred either because mustard oil was deliberately adulterated with a cheaper substitute or because the seeds of the prickly poppy, *Argemone mexicana*, closely resembled those of the mustard plant, with which they became mixed during harvesting and milling.[138] Here was another example of how the language of toxicity and the scientific investigation of its nature and effects were shifting away from old concerns – datura *thugi*, arsenic murders, cattle poisoning – to a newer range of poison substances and agencies, such as contaminated milk and adulterated cooking oil, that were purchased in the marketplace or delivered to the home.

Many of the developments discussed in this chapter had their origins elsewhere or exhibited close parallels with events and discoveries in other parts of the world. The genealogy of the Indian Poisons Act of 1904 can be traced back to the British 'arsenic panic' of the 1840s and to subsequent metropolitan enquiries and legislative enactments, just as the importation of arsenic into nineteenth-century India reflected the expansion and changing nature of international trade. The profound changes in medical thought and sanitary practice that followed the rise of bacteriology and the creation of a newly interventionist public health ethos likewise had their origins elsewhere, in Europe. Even the groundbreaking investigation into beriberi owed more to research carried out in Java, Malaya and the Philippines than it did to any comparable work in India. In some ways, British India was remarkably slow to act in relation to all these diverse manifestations of 'poison'. Many factors sustained this inactivity – the conservatism of the colonial medical establishment, the regime's pragmatic post-1857 policy of non-intervention and laissez-faire, and the preoccupation with other issues (such as famine and malaria) which seemed far more costly and urgent than the relatively small number of arsenic poisonings and suspicious cattle deaths. Perhaps, there was even a kind of underlying fatalism attached to the colonial view of poisoning in India as being too culturally pervasive, too environmentally embedded, to be responsive to any affordable measure of containment or extirpation.

But many things changed in the 1890s and 1900s, as science and medicine in India belatedly caught up with developments across the

---

[138] R. B. Lal and S. C. Roy, 'Investigations into the Epidemiology of Epidemic Dropsy, Part I', *IJMR* 25 (1937–38): 163–76; R. B. Lal and A. C. Das Gupta, 'Investigations into Epidemic Dropsy, Part X', *IJMR* 29 (1941): 157–65.

Western world – or even, as the Secretary of State saw it, as the anomaly of India's backwardness among civilized nations was at last being addressed and at least partially resolved. The argument, often expressed, that India was 'different' – and therefore dangerous – did not thereby disappear. Rather, it was resurrected in new forms. Datura *thugi* and cattle poisoning might no longer define the Oriental, but through the science of tropical medicine and disease, and through the continuing emphasis on poisons as both dangerous and curative, that sense of difference (and therefore danger) was kept alive. But, as this chapter has also tried to suggest, one of the things that changed in the 1890s and 1900s was the growing sense that Indians had their own interest in the control of poisons, from unregulated arsenic in the bazaars to adulterated milk and contaminated cooking oil in the home.

Chunilal Bose's call for an Indian poisons act was a clear demonstration of this emerging sense of Indian ownership over the problem of poisons. Just as some Europeans had come by the early twentieth century to think of themselves and their households as vulnerable to poisoning by Indian servants and miscreants, so still more acutely middle-class India and *bhadralok* Bengal thought of themselves as liable to accidental poisoning by the low-caste street vendors and servants who supplied them – and thereby poisoned them – with adulterated cooking oil or bacteria-laden ice cream. Even the fact that cow killing by means of arsenic appeared to be the work of Chamars, among the lowest of Hindu castes, strengthened the association of poisoning with the lower orders and their uneducated ways. The well-being, the intimate appetites and public desires of the higher castes and middle classes were as much at stake as those of Europeans. This poison fear had particular resonance in Bengal, where the Hindu elite already felt itself besieged and on the verge of becoming a 'dying race' – as it struggled to keep pace with the growing numbers of Muslims and lower-caste poor, as malaria and kala-azar ravaged the countryside and as its most talented individuals fell prey to diabetes, tuberculosis or suicide.[139] Literally or figuratively, poisoning seemed to encapsulate their collective vulnerability and, unless science could save them, warned of their imminent fate.

---

[139] On the 'dying race' argument, see Papia Chakravarty, *Hindu Response to Nationalist Ferment* (Calcutta: Subarnarekha, 1992), 30–53.

7   Polluted places, poisoned lives

Nowhere was poison's widening penumbra more evident, and its toxic association with contamination and pollution more apparent, than in India's cities. Cities were sites of animal as well as human concentration. They were places of poverty and slums, of tainted water supplies and adulterated food. Cities spawned 'offensive' trades and public 'nuisances', industries and engines that fouled the air and water around them, factories and workshops that posed health hazards for those who worked in them as much as for those who inhabited adjacent neighbourhoods. In the city, still more than in the countryside, especially by the 1890s and 1900s, the 'social life' of poisons intersected with environmental history and with environmental governance.[1] Cities were primary locations for attempts not only to regulate and manage toxicity, to observe its malignant presence and bring it under scientific scrutiny and control, but also to harness the power of poison in pursuit of public health and the public good. Cities were centres of a vocal Indian as well as European public and the articulation of toxic concerns grounded as much in class identities as in colonial anxieties and anti-colonial antipathies. More than anywhere else, cities combined the almost everyday experience of toxicity with an effusion of toxic imagining that at times bordered on an apocalyptic 'fear of a poisoned world'. The toxic history of Indian cities reached back into the early nineteenth century (and beyond) but it also foreshadowed many aspects of India's toxic future. The 'contaminated city' of the colonial era serves as a precursor and proxy for toxicity in the post-colony.[2]

---

[1] I prefer 'environmental governance' to the more Foucauldian 'environmental governmentality' employed by others: for example, Kelly D. Alley, *On the Banks of the Ganga: When Wastewater Meets a Sacred River* (Ann Arbor: University of Michigan Press, 2002), 22.

[2] Colin McFarlane, 'Governing the Contaminated City: Infrastructure and Sanitation in Colonial and Post-Colonial Bombay', *International Journal of Urban and Regional Research* 32 (2008): 415–35.

Animal cities

The cities of British India were home to animals, not just people – animals that provided humans with food and drink, served the needs of transport and industry and lived on human profligacy and waste. Animal geography helped define urban spatiality and shape practices of urban inclusion and exclusion, but it also marked out a critical terrain for toxic discourse and the quest for remedial action. Bombay and Calcutta, India's most populous cities, each had human populations close to 1 million by 1914, but it is much harder to ascertain the size of their animal populations and to assess the impact animals had on human health and urban governance.[3] However, the Bombay Municipal Act of 1865 established a new regime of urban accountability, and the reports of the municipal health and sanitary officers give some insight into the city's animal life.[4] Large numbers of cows and buffaloes were kept in the city and provided its main supply of milk. In 1873, when the human inhabitants numbered just over 665,000, there were an estimated 1,972 milk-producing animals in Bombay, with two crowded inner-city wards alone accounting for nearly 800 of these. Since cows, pigs and other productive animals were mostly kept by poor households, high animal densities tended to correlate to concentrations of urban poverty and slum dwelling.[5] As well as being protected by their status as sacred animals, cows were saved from urban exclusion by their utility as a provider of income (and, to some degree, nutrition) for the poor and as a source of cow dung fuel. Despite growing importation of milk from the countryside, in 1924 there were still 93 licensed cattle byres in the city, housing an estimated 5,000 animals and yielding 18,000 gallons of milk a day.[6]

Materially and perceptually, the well-being of Bombay's human inhabitants was closely bound up with its animal population and the 'nuisance' their presence created. 'Nuisance' was a term widely employed in the discourse of urban governance in nineteenth- and early-twentieth-century India. Technically, it referred to any 'act, omission, place or thing' which caused 'injury, danger, annoyance or offence to the sense of sight, smell or hearing' or which was 'dangerous to life or injurious to health or property'.[7] Although 'nuisance' at times might signify something that was irritating or irksome to the senses rather than being actively dangerous, the term could also convey miasmatic associations with pollution, disease and poison.

---

[3] In 1911, Calcutta had a population of 1,013,143 and Bombay 979,445. The next largest city was Madras with 518,660.
[4] RHO, *ARMC (Bombay), 1866*, 1.   [5] RHO, *ARMC (Bombay), 1873*, 76–83.
[6] RHO, *ARMC (Bombay), 1923–24*, 40.
[7] A. J. H. Russell (ed.), *McNally's Sanitary Handbook for India* (6th ed., Madras: Superintendent, Government Press, 1923), 449.

Valued though their contribution to urban life was in other respects, animals were seen as a potential danger to human health and an intrusion into the modern city of forms of nature better suited to field and farm. Cattle were crowded into small, rickety sheds, almost devoid of light, ventilation and drainage, while the owners lived in squalor above or beside their beasts. In contemporary sanitary thought, animal pollution was intimately connected to contagious disease and miasmatic poison. Bombay's first Health Officer, T. G. Hewlett, wrote in 1866: 'The heat, the faint sickening odour, the walls moist with the exhalation from the animals, the stifling smell of ammonia, make one sick on first entering the stables ... an atmosphere impregnated with such impurities must be unhealthy.'[8] Cattle could not be excluded from the city, but successive health officers in Bombay sought to control the stabling of animals and enforce cleanliness in the surrounding yards. All cattle sheds had to be licensed and open for inspection, with fines for non-compliance.[9]

The authorities waged a constant war against what were described as 'dangerous' and 'offensive' trades, many of which had animal connections. These included slaughterhouses, tanneries, warehouses for storing dried fish and yards for making catgut from animal intestines.[10] In 1867, the slaughter of cows, sheep and goats was banned from central Bombay and relocated to the city's northern perimeter. Butchers fiercely opposed the move: they went on strike for a week but soon succumbed.[11]

Animals were urban scavengers, with pigs, dogs and vultures almost constantly in evidence. In the 1870s, by-laws were introduced in Bombay to curb the 'pig nuisance', requiring that the animals be kept well away from human habitation.[12] Each year, large numbers of dead animals were collected from the streets, taken to the outskirts of the city and dumped. In 1872, more than 5,000 animal carcasses were disposed of in this way – or, to reproduce the almost surreal precision of the municipal reports, 188 buffaloes, 593 bullocks, 608 calves, 733 sheep, 491 goats, 213 horses and 2,725 dogs.[13] Like snakes in the countryside, dogs were a cause of particular concern.[14] Part of the reason for this was fear of rabies, a disease that caused thousands of deaths a year in India.[15] But dogs, left

[8] RHO, *ARMC (Bombay), 1866*, 4.
[9] RHO, *ARMC (Bombay), 1873*, 76–83; RHO, *ARMC (Bombay), 1880*, 311–12.
[10] RHO, *ARMC (Bombay), 1866*, 25.   [11] RHO, *ARMC (Bombay), 1867*, 13.
[12] RHO, *ARMC (Bombay), 1879*, 280–81.   [13] RHO, *ARMC (Bombay), 1872*, 13.
[14] In Bombay between 1826 and 1832, an estimated 63,000 dogs were destroyed: Jesse S. Palsetia, 'Mad Dogs and Parsis: The Bombay Dog Riots of 1832', *Journal of the Royal Asiatic Society* 11 (2001): 26.
[15] On rabies and its treatment, see Pratik Chakrabarti, *Bacteriology in British India: Laboratory Medicine and the Tropics* (Rochester, NY: University of Rochester Press, 2012), 66–71, 93–94.

to wander, scavenge, defecate and die in the streets, were seen more generally as a form of urban pollution and a threat to public health. In 1873, 5,988 dogs were destroyed by Bombay's municipal authorities. Fifty years later, in 1923, 10,617 dangerous or stray dogs were caught. Of these, only ten were returned to their owners: the rest were poisoned with strychnine or gassed in the municipal 'lethal chamber'.[16] Toxicity thus entered urban sanitary practice as a solution to a problem of animal excess, not just as an environmental hazard to be surmounted.

More dramatically still, the epidemic of bubonic plague that erupted in Bombay in 1896 and then surged across India, resulting in more than 8 million deaths by 1914, stoked fresh alarm about animal, this time rodent, populations. Initial attempts to control the epidemic encountered serious difficulties, including popular resistance to body inspections, the segregation and hospitalization of plague suspects and the plague camps to which evicted residents were sent.[17] The anti-plague serum developed by Waldemar Haffkine in Bombay also attracted opposition and revived rumours about imperious Britons intent on poisoning Indians. An incident in Punjab in November 1902, in which nineteen villagers died after being inoculated with Haffkine's serum, fuelled further resistance. The episode also increased scepticism among colonial officials and sanitarians who already doubted that vaccination was an adequate solution to the escalating plague problem.[18] Rat control suggested a less contentious alternative, despite the fact that many Hindus and Jains were opposed to the taking of animal life. Rat eradication became even more appealing once medical experts in India came to recognize around 1905 that rats' fleas were instrumental in the transmission of the plague bacillus from rodents to human hosts.[19] The destruction of rats became for the Government of India 'the most important preventive measure of all', though it was left to the local authorities to decide how best to implement this policy.[20] Small

---

[16] RHO, *ARMC (Bombay), 1873*, 69; RHO, *ARMC (Bombay), 1923–24*, 40. In the mid-1930s, the 'lethal chamber' was replaced by electrocution: RHO, *ARMC (Bombay), 1934–35*, 46.

[17] John T. W. Leslie, Sanitary Commissioner with the Government of India to Secretary, India (Home), 31 October 1905, India, Home (Sanitary), nos 327–47, February 1906, NAI; RHO, *ARMC (Bombay), 1896–97*, 661–779. Also, David Arnold, *Colonizing the Body: State Medicine and Epidemic Disease in Nineteenth-Century India* (Berkeley: University of California Press, 1993), ch. 5; Raj Chandavarkar, 'Plague, Panic and Epidemic Politics in India, 1896–1914', in Terence Ranger and Paul Slack (eds), *Epidemics and Ideas: Essays on the Historical Perception of Pestilence* (Cambridge: Cambridge University Press, 1992), 203–40.

[18] For attitudes to plague control, see Chakrabarti, *Bacteriology*, 55–58.

[19] W. G. Liston, 'Plague, Rats, and Fleas', *IMG* 40 (1905): 43–49.

[20] Leslie to Secretary, India, Home (Sanitary), nos 327–47, February 1906, NAI. In 1914, Punjab's Sanitary Commissioner stated that rat poisoning had saved more lives in the province than any other plague control measure: S. Browning Smith, 'Rat Destruction

sums of money were paid for each rat destroyed, just as rewards had earlier been given for killing venomous snakes. Poisoning might, as the name of one proprietary brand of rat poison indicated, be 'Rough on Rats', but, as a technique of disease control, it appeared to work, and in towns and cities as well as in the countryside.[21] Various forms of poisoned bait were used. Of the chemical poisons, barium carbonate was considered the cheapest and most effective, but experiments were conducted to ascertain other ways of overcoming 'the rat problem'.[22]

Among the solutions most widely advocated was the use of poisonous hydrocyanic acid gas, pumped into grain stores, warehouses, tenement blocks and docked ships to exterminate both rats and their fleas. Trials with this and other toxic gases began in 1915 and continued into the 1930s, with the Haffkine Institute in Bombay taking the lead in poison gas research as it did in venom and vaccine trials.[23] In contrast to international condemnation of the military use of poison gas during the First World War, its deployment against rodents and insects commanded strong scientific support. In 1919, Glen Liston, IMS, anticipated the coming of the 'next war' – the war of 'Man versus Insects' – in which toxic gas would be essential in destroying insect pests and vectors. Two-thirds of preventable disease in India was due to insects, he told the Indian Science Congress, and the cost to the economy, as well as to public health, was enormous.[24] Especially in cities, hydrocyanic gas offered an effective solution to this problem: 30 units of the gas could fill 100,000 units of air within two hours and rid the entire space of all insects present. It could even penetrate the spaces between bales and sacks where insects tended to hide.[25] Hydrocyanic gas was, according to Liston, safe and easy to use so long as 'a moderate amount of caution' was taken. Accidents might happen, but humans, reassuringly, were 'less susceptible to the poison' than animals.[26]

and Plague', *Supplement to the Indian Journal of Medical Research: Proceedings of the Third All-India Sanitary Conference* (5 vols, Calcutta: Thacker, Spink, 1914), 5: 158–61.

[21] RHO, *ARMC (Bombay), 1904–05*, 196.

[22] J. G. C. Kunhardt, 'The Rat Problem in India', *IJMR*, special issue (1919), 145–72. On experiments with rat destruction, see India, Home (Sanitary), nos 329–45, 7 November 1905, NAI; RHO, *ARMC (Bombay), 1908–09*, 31.

[23] W. Liston, 'The Use of Hydrocyanic Gas for Fumigation', *IJMR* 7 (1919–20): 778–802; S. N. Gore, 'Calcium Cyanide Fumigation', *IJMR* 13 (1925–26): 287–99; S. S. Sokhey, G. D. Chitre and S. K. Gokhale, 'The Relative Value of Some Proprietary Cyanide Preparations for the Extermination of Rats and Fleas as a Plague-Prevention Measure', *IJMR* 27 (1939–40): 389–407.

[24] W. Glen Liston, ' "The Next War": Man *versus* Insects', *IJMR*, special issue (1919), 18–25.

[25] W. Glen Liston and S. N. Gore, 'Abstract of a Paper on Hydrocyanic Acid Gas as an Insecticide', ibid., 40–42.

[26] Liston, 'Hydrocyanic Gas', 781, 793. The dangers of hydrocyanic acid were, however, apparent when a researcher at the Bengal Chemical and Pharmaceutical Works in

Figure 7. 'Bombay City: Rat Examination at the Laboratory', *Journal of Hygiene* 7, 1907, plate 20.

Poison was thus assigned a new utility in the war against tropical disease. The number of rodents destroyed by poisoning and other means is staggering. Between 1897 and the end of January 1902, 797,386 rats were slaughtered in Bombay, a number almost equal to the city's human population.[27] As scientific conviction grew that rats and their fleas spread plague, the figure soared, reaching more than 650,000 in 1913–14. Each year, thousands of rats were sent to the Haffkine Institute to be examined for signs of plague (see Figure 7). In 1905, 53,096 rats were sent there: of these 42,142 were examined and 838 showed signs of infection. In 1914–15, 210,080 out of the 644,241 rats collected were sent for examination and of these 12,530 were infected.[28] Human plague deaths in the city peaked in 1903, but the number of rats trapped and killed continued to rise. In 1916, 943,346

Calcutta was accidentally poisoned by the gas and died: P. C. Ray, *Life and Experiences of a Bengali Chemist* (Calcutta: Chuckerbutty, Chatterjee, 1932), 107.

[27] RHO, *ARMC (Bombay), 1904–05*, 196.

[28] RHO, *ARMC (Bombay), 1905–06*, 202–03; RHO, *ARMC (Bombay), 1914–15*, 32.

rats were destroyed.[29] Between 1897 and the mid-1920s, close to 15 million rats were killed in Bombay, many of them by means of poison.[30]

The polluted city

The discourse of urban toxicity in India was complicated by the variable meaning ascribed to the term 'pollution'. In much of the colonial literature, 'pollution' served, as in contemporary Britain, as an environmental concept even before modern ideas of 'the environment' had been developed.[31] It signified the unwanted, and almost certainly unhealthy, contamination of air, water and soil by human waste and animal detritus, and by the dust, smoke, sludge and effluvia that issued from chimneys, drains and sewers or emanated from slums, stables, cowsheds, workshops and factories. One illustration of this sanitary notion of pollution, and its continuity with ideas of urban 'nuisance' and miasmatic 'poison', concerns disposal of the human (rather than animal) dead. In the 1850s, Bombay's health authorities expressed alarm at the custom attributed to Hindu castes of burying the infant dead and the foetuses of stillborn babies in back gardens and courtyards – what A. H. Leith referred to as 'the clandestine domiciliary internment of infants'. The concealment of corpses anyway created suspicion that this might represent the 'undetected commission of infanticide'. But secretive burials were also deemed 'prejudicial to the public health and welfare'. Stillborn and infant bodies were also found half-buried in the sands of Bombay's Back Bay. Although these human remains decomposed rapidly in the wet ground, they were still thought to generate 'poisonous gases' and to pollute the air of nearby dwellings, one indication among many of how frequently ideas of poison and pollution overlapped and intersected in contemporary sanitary discourse.[32]

In the 1850s, Bombay had more than 60 cemeteries and cremation grounds scattered across the city.[33] Cremation, though widely practised

---

[29] RHO, *ARMC (Bombay), 1916–17*, 29.

[30] Rat destruction continued into the 1930s, with more than 700,000 rodents killed annually: *ARMC (Bombay), 1934–35*, 19.

[31] For example, Lawrence E. Breeze, *The British Experience with River Pollution, 1865–1876* (New York: Peter Lang, 1993). For the more recent understanding, see David Briggs, 'Environmental Pollution and the Global Burden of Disease', in David J. Briggs, Michael Joffe and Paul Elliott (eds), *Impact of Environmental Pollution on Health: Balancing Risk* (Oxford: Oxford University Press, 2003), 1–24; Marquita K. Hill, *Understanding Environmental Pollution* (Cambridge: Cambridge University Press, 1997).

[32] *Correspondence Relating to the Prohibition of Burials in the Back Bay Sands* (Bombay: Bombay Education Society's Press 1855), 9; *Correspondence Relating to a Proposed Enactment for the Regulation of Places Used for the Disposal of Corpses in the Town and Island of Bombay* (Bombay: Bombay Education Society's Press, 1855), 11–12.

[33] *Correspondence Relating to the Prohibition of Burials*, 20–22.

among Hindus, was not necessarily the norm. In 1873, according to official statistics, 4,786 corpses were cremated compared to 10,693 buried and 971 exposed on the Parsi Towers of Silence.[34] Where bodies were interred, they were often buried too close to the surface and so exposed to the ravages of dogs and jackals. After generations of use, inner-city cemeteries had become overcrowded and, in releasing supposedly poisonous gases and dangerous effluvia into the air, threatened to be 'injurious to the health of the population dwelling around them'.[35] Three Jewish burial grounds on Grant Road were scheduled for removal in 1867 because 'the exhalations from them cannot but exert a prejudicial influence on the health of the people who live in the neighbourhood'.[36] Anxiety about effluvia leaking from burial grounds prompted some sanitarians to argue that cremation (still regarded with some horror in Britain) was a preferable method of disposing of the dead in India – despite the destruction of potentially incriminating forensic evidence – since it would destroy 'disease germs' and eliminate the repulsive smell of decomposing corpses.[37] But, in this sensory world of sanitation, in which nuisance, disease and poison were so closely allied, the foul smoke from cremation sites and burning ghats, drifting across residential areas of the city, also drew strong protests, especially from Europeans. The dead were a constant trouble to the living.

The real or alleged toxicity of air, food and water underscored the seemingly perilous nature of urban living. At times, the reference to 'poison' was little more a gesture of distaste and repugnance, as when Calcutta's *Englishman* newspaper remarked, in complaining about the stench from city sewers: 'This smell is lethal.'[38] But at other times the danger posed by urban poison was a material reality. In 1872, three workers at the Howrah gas plant died from exposure to a build-up of coal gas, and on at least two occasions municipal workers in Bombay, sent to inspect drains or unblock sewers, were overwhelmed by 'mephitic vapours' and died.[39] On other occasions again, the connection was more speculative. In 1885, a European doctor informed a specially convened sanitary commission in Calcutta that members of the public were

---

[34] RHO, *ARMC (Bombay), 1873*, 93.

[35] RHO, *ARMC (Bombay), 1872*, 15; Hehir, *Hygiene*, 408.

[36] RHO, *ARMC (Bombay), 1867*, 5.

[37] Patrick Hehir, *Hygiene and Diseases of India: A Popular Handbook* (3rd ed., Madras: Higginbothams, 1913), 407–09.

[38] *Report of the Commission Appointed to Enquire into Certain Matters Connected with the Sanitation of the Town of Calcutta* (Calcutta: Bengal Secretariat Press, 1885), appendix 2, lxvii.

[39] D. B. Smith, 'Deaths from Coal Gas', *IMG* 7 (1872): 76–77; RHO, *ARMC (Bombay), 1895–96*, 583; RHO, *ARMC (Bombay), 1907–08*, 141.

being made ill as a result of poisonous gases escaping from municipal sewers and leaking into their homes.[40] Rumours of men suddenly suffocated by 'mephitic vapours' and reports of 'rapid poisoning through the foul air of sewers' became rife too as Bombay grappled with bubonic plague in 1896.[41] What Lawrence Buell has called 'fear of a poisoned world' was widespread in urban India, even among sanitary officials.[42]

In British India, pollution had a secular, scientific and proto-environmental meaning that coexisted, often uneasily, with the concept of pollution in Hindu ritual and social practice, though both were united in having strong moral and spatial, as well as corporeal, connotations.[43] Ritually, pollution defined the prescribed distance between different castes and the impure state into which men and women lapsed through contact with human and animal waste and association with contaminating phases of the human life cycle from birth and menstruation to the corruption of the dead. Pollution could be communicated through the food, water and other objects which a person of lower caste might have handled or used.[44] Ritual pollution further encompassed what were seen as degrading occupations, especially the polluted and polluting trades followed by untouchable leather workers or municipal sweepers like the *halalkhors* of Bombay.[45] For such social groups, and those who sought to avoid them, pollution constituted a kind of social poison. In this way, ritual pollution tended to reinforce the association of material pollution – even poisoning – with poor and low-status communities.

In 1847, the engineer F. W. Simms, addressing the problem of Calcutta's water supply, remarked that high-caste Hindus opposed the use of water that had been polluted (in a religious sense) by passing through iron pipes. Instead, they continued to draw water from the Hooghly (a distributary of the sacred Ganges), which they considered ritually pure. But he believed that once 'the first shock to the prejudices of the orthodox Hindus' had passed, they would welcome access to a water supply that would also escape pollution (in a sanitary sense) caused 'either

[40] Dr G. R. Ferris in *Report of the Commission*, appendix 2, lxii.
[41] RHO, *ARMC (Bombay), 1896–97*, 593.
[42] Lawrence Buell, 'Toxic Discourse', *Critical Enquiry* 24 (1998), 640.
[43] On this distinction, see Awadhendra Sharan, *In the City, Out of Place: Nuisance, Pollution, and Dwelling in Delhi, c. 1850–2000* (New Delhi: Oxford University Press, 2014), 29.
[44] Edward B. Harper, 'Ritual Pollution as an Integrator of Caste and Religion', *Journal of Asian Studies* 23 (1964): 151–97; Louis Dumont, *Homo Hierarchicus: The Caste System and Its Implications* (revised ed., Chicago: University of Chicago Press, 1979), ch. 6.
[45] Harold H. Mann, 'The Untouchables of an Indian City', in Daniel Thorner (ed.), *Harold H. Mann: The Social Framework of Agriculture: India, Middle East, England* (Bombay: Vora, 1967), 175–91; Mary Searle-Chatterjee, 'The Polluted Identity of Work: A Study of Benares Sweepers', in Sandra Wallmann (ed.), *Social Anthropology of Work* (London: Academic Press, 1979), 269–86.

by accident or design'. The municipal water supply would be safer to drink and so would ultimately prove more acceptable. High-caste Hindus had also objected, Simms explained, to the municipal supply because the engine that pumped water from the Hooghly was greased with (ritually polluting) animal tallow, but putting a Brahmin in charge of the pumping works had reconciled the higher castes to its use. He concluded that the 'prejudices' of the orthodox Hindus would disappear 'when they find that they can obtain close to, or in their houses, the sacred fluid of the Ganges, drawn ... from above the pollutions of the City'.[46] In other words, both ritual and environmental pollution would thereby be avoided.

Given the social as well as environmental issues that surrounded it, the problem of water – pure or polluted – was not easily resolved. Indeed, in the view of colonial sanitarians, it presented difficulties beyond those encountered in contemporary Europe. 'To anyone familiar with this country', observed J. A. Jones, a sanitary engineer in Madras in 1896, 'this pollution [of water] is daily brought within their observation; people habitually defile the slopes and margins of tanks, and then cleanse them- selves in the water'.[47] He claimed that this pollution, or 'poisoning', was not just a negation of Western sanitary ideals but also counter to the Laws of Manu and other ancient Hindu authorities.[48] But environmental as well as social factors were also invoked – the monsoon climate left water- courses dried up or stagnant for much of the year, then flooded to over- flowing during the rainy season. The heat and humidity of 'tropical' India added still further to an 'Oriental' or 'Eastern' urban environment – to the abundance of flies and the stinking putrefaction of human, animal and vegetable waste. Even the primitive means used to collect human excreta in cities like Bombay – by means of basket-carrying municipal sweepers and the vile stench as night soil carts trundled through the streets – created conditions unlike a modern, Western, sanitized city.[49]

It could be argued that these prejudicial observations conveniently deflected responsibility for the many shortcomings and failures in muni- cipal waste management and water provision onto local environmental and indigenous social conditions, and, thereby, in making the problems

---

[46] F. W. Simms, *Report on the Establishment of Water-Works to Supply the City of Calcutta* (Calcutta: Military Orphan Press, 1853), 25–26.

[47] J. A. Jones, *A Manual of Hygiene: A Manual of Hygiene, Sanitation and Sanitary Engineering with Special Reference to Indian Conditions* (Madras: Superintendent, Government Press, 1896), 12. On similar issues in Delhi, see Michael Mann, 'Delhi's Belly: On the Management of Water, Sewage and Excreta in a Changing Urban Environment during the Nineteenth Century', *Studies in History* 23 (2007): 1–31.

[48] Jones, *Manual*, 12–13; *McNally's*, 36.

[49] William Wesley Clemensha, *Sewage Disposal in the Tropics* (Calcutta: Thacker, Spink, 1910), 1; J. A. Turner, *Sanitation in India* (Bombay: 'Times of India', 1914), 1, 171.

appear entrenched and insurmountable, contributed to a state of sanitary stasis. But, in reality, repeated interventions were made by the municipal authorities in an attempt to improve the supply of potable water and to restrict access to polluted sources. Repeated efforts were made, as in Bombay, to fill in tanks and cover over wells that were pronounced unfit for use, despite the fact that these sources of water held great religious significance for Hindus and Parsis. Concepts of what constituted purity and pollution might often collide: a well could provide water that was ritually pure but which, seen under the microscope, teemed with bacilli.[50]

The attempt to restrict, and even eliminate, existing water sources like tanks and wells made it all the more imperative that the municipal authorities follow British precedent and create their own, uncontaminated, supplies of potable water. In order to do this, water had to be either extracted from rivers, like the Hooghly in Calcutta, and filtered through pumping stations, or, as in Bombay, collected and stored in reservoirs, sourced from monsoon rain and then distributed through pipes to residents across the city. Both presented serious difficulties. As Calcutta expanded, and as the banks upstream from the city became flanked by jute mills, paper works, distilleries and other factories discharging industrial effluent and untreated excrement into the river, so the danger grew that the water extracted from the river would prove chemically and bacteriologically unfit to drink, and would further raise doubts about its ritual purity. In 1919, a Hooghly River Pollution Inquiry was set up, with one bacteriological and two chemical assistants, to investigate the quality of the city's water. It became clear that apart from offending factories, there were dozens of sites along the bank where municipal drains emptied into the river or where human excreta exposed the river to the dangers of 'constant pollution'.[51] One of the tasks of the municipal laboratories was to use their scientific expertise not just to propose means (such as septic tanks) to reduce the flow of pollution into the river but also to persuade a doubtful public that the water supplied to consumers was safe to drink and as pure as possible. Assuming their reliability, the statistics were impressive. Until the 1880s, analysts had only limited means of determining pollution levels – by observing the colour, murkiness and odour of the water and analysing its chemical content. But the advent of bacteriology made the testing and purification of water appear more scientifically reliable. It now became possible to identify and measure bacterial as well as chemical content. In 1896, Jones confidently

[50] RHO, *ARMC (Bombay), 1912*, 38.
[51] *Report on Sanitation in Bengal, 1919*, 24; *Annual Report of the Director of Public Health, Bengal, 1920*, 22.

claimed that between 50 and 100 per cent of the solids in water extracted from the Hooghly were removed by the filtering process, which then 'delivered clear and transparent water to the consumers in Calcutta'.[52]

Opinion about the purity (or otherwise) of India's rivers was contradictory. Even in the 1890s, some commentators held that river pollution on the British scale was a remote possibility in India, given how far it lagged far behind in urbanization and industrialization.[53] This view drew support, too, from the Hindu belief that flowing water purified itself and that the holy Ganges was able to cleanse itself of any impurities. E. H. Hankin, UP's Chemical Examiner and Bacteriologist, seemed to give scientific endorsement to this claim in 1894 by arguing that the river had a 'germicidal action' that destroyed harmful microbes.[54] This proposition was quickly challenged as encouraging the belief that sewage and other waste could be dumped into Indian rivers without any harmful effects.[55] A correspondent of the *Lancet* expressed similar incredulity, remarking that 'that most sacred river, the Ganges, is a vast sewer receiving the sewage of all the sewered towns on its banks'. He poured scorn on claims that the river was able to cleanse itself as 'disastrous to public health'.[56] That writer, and others, cited Cawnpore (Kanpur) as a major source of river pollution. Its textile mills and tanneries then – as they still do now, but in vastly greater quantities and to far greater toxic effect – spewed human detritus, industrial waste and chemical effluent into the Ganges in an ever increasing tide.[57]

The task of providing pure water and countering pollution in Bombay proved even more exacting than in Calcutta. In the 1870s, it was said that many inhabitants were drinking water from wells and tanks that was 'semi-liquid mud' and 'could only be regarded as sewage'.[58] Municipal action seemed to compound the problem, as much of the city's night soil was dumped untreated into the harbour – prompting complaints from visiting ships that the surrounding air and water was 'thick with pollution'

[52] Jones, *Manual*, 15, 17–24.
[53] Editorial, 'River Pollution in India', *IMG* 25 (1890): 56–57; Mann, 'Delhi's Belly', 12.
[54] E. H. Hankin, 'Microbes of Indian Rivers', *British Medical Journal* 1780 (1895): 312. Cf. David L. Haberman, *River of Love in an Age of Pollution: The Yamuna River of Northern India* (Berkeley: University of California Press, 2006); Meenakshi Sharma, 'Polluted River or Goddess and Saviour? The Ganga in the Discourses of Modernity and Hinduism', in Helen Tiffin (ed.), *Five Emus to the King of Siam: Environment and Empire* (Amsterdam: Rodopi, 2007), 31–50.
[55] [Anon], 'The First Indian Medical Congress', *British Medical Journal*, 9 February 1895, 310–13.
[56] [Anon.], 'Notes from India', *Lancet*, 2 September 1911, 731.
[57] Ibid., 731; Ahmad Mukhtar, *Report on Labour Conditions in Tanneries and Leather Goods Factories* (Simla: Manager, Government of India Press, 1946), 20–21.
[58] Hector Tulloch, *The Water-Supply of Bombay* (London: W. J. Johnson, 1872), 211.

and crews were being 'poisoned' by water taken from the harbour.[59] In order to provide Bombay with a less polluted supply, reservoirs were constructed to the north of the city. When the first of these, at Vehar, opened in 1860, the water pumped into city homes was itself prone to pollution, either at source, in the reservoir, or in the course of transmission through contamination from soil and sewage. It was said in 1872 to be 'a shame' that the people of Bombay were 'still in too many instances allowed to poison themselves' by drinking water which was 'only diluted sewage'.[60] Three years later, the city's Health Officer was still voicing concern. 'Considering the very great facilities that exist for the pollution of the water in the Vehar dipping-wells', he reported, 'we need not be surprised when we happen to find water in them so impure as to be not only unfit for drinking, but absolutely poisonous'.[61] When a second reservoir, the Tulsi, began supplying the city in 1879, its water, too, appeared unfit to drink. Thick with 'organic pollution' of a dirty green colour, it 'absolutely stank'. Improvements were made, but concern about the quality of water from the two reservoirs persisted for years. Only gradually did the quality improve and public protest decline.[62] And yet, despite this, colonial sanitarians remained confident that piped water would eventually result in the elimination of diseases like cholera and typhoid which had waterborne pollution as their 'chief, if not only, exciting cause'.[63]

As in Calcutta, by 1900, it was possible to present the Bombay public with a wealth of technical and statistical data to demonstrate the purity of the municipal water supply. Filtering, for instance, was said to remove 89 per cent of bacteria from water supplied to the city. This assertion had a particular relevance as in 1900 more than 4,000 people died during a cholera epidemic. Daily bacteriological examinations were made of water in the three municipal reservoirs by then in operation. No evidence of the cholera bacillus was found, suggesting that the outbreak was not caused by the municipal water supply but by remaining wells and tanks.[64] In 1905, of the 1,701 items examined in Bombay's municipal laboratories,

[59] RHO, *ARMC (Bombay)*, *1875*, 138–39. Similar complaints were made about Calcutta's harbour becoming a 'maelstrom of death' from sewage dumped in the Hooghly: Editorial, 'Cholera in the Port', *IMG* 1 (1866): 191.

[60] RHO, *ARMC (Bombay)*, *1872*, 24. On the issues surrounding this project, see Mariam Dossal, 'Henry Conybeare and the Politics of Centralised Water Supply in Mid-Nineteenth Century Bombay', *Indian Economic and Social History Review* 25 (1988): 79–96.

[61] RHO, *ARMC (Bombay)*, *1875*, 145.

[62] RHO, *ARMC (Bombay)*, *1879*, 276; RHO, *ARMC (Bombay)*, *1894–95*, 753.

[63] T. Blaney, 'Brief Notes on Enteric Fever', *IAMS* 10 (1870): 357.

[64] C. H. Cayley, 'Report on the Analyses of the Bombay Water-Supply', in RHO, *ARMC (Bombay)*, *1900–01*, 492, 494; RHO, *ARMC (Bombay)*, *1901–02*, 171.

763 related to the municipal water supply and 273 to tank and well water. Few of the piped-water samples suggested anything alarming.[65] Science thus had a vital role in detecting pollution (and the 'poison' identified with it), but also in reassuring an apprehensive public. In practice, though, the work of the bacteriologist and the chemical analyst-examiner could not alone resolve the monumental problems of health and sanitation afflicting the city.[66]

Air as well as water stood high on the urban environmental agenda. According to J. A. Jones in 1896, the air of Indian cities was 'dangerously impure'. It was 'contaminated by the poison coming from the soil fouled by the decomposing excreta of the present and past generations'. Much the same, he noted, had once been true of London and Edinburgh, but that was back in the seventeenth century: India had two centuries of sanitary improvement to catch up on.[67] One of the earliest attempts to address atmospheric pollution was through measures to regulate 'dangerous and offensive trades'. Many of these were small-scale enterprises, making catgut, boiling tallow for candles, drying fish or crushing animal bones, but others, including lime kilns, brickworks, tanneries and dyeing and printing works, might involve substantially larger workforces. As early as the 1830s, magistrates in Bombay sought to expel the worst of these and relocate them outside city limits. There and in Calcutta, initial pressure for the removal of 'offensive trades' often came from governors and senior officials, who expressed personal objections to such 'nuisances'.[68] By the late nineteenth century, the exclusion of such trades from urban areas – what would now be called 'zoning' – had become a standard recommendation in India's sanitary textbooks. In 1913, still hovering between the old language of miasmas and the new bacteriology, Patrick Hehir remarked that almost all 'offensive trades', from piggeries to tanneries, owed their 'unpleasantness and injurious effect to the decomposition of animal matter'. They produced 'foul-smelling gases and bacteria' that blighted entire neighbourhoods. 'No one', he declared, 'has a right to contaminate the air breathed by his neighbours'.[69]

Objections came, too, from residents, Indian as well as European, who opposed the locating of latrines, slaughterhouses or bone-crushing plants in their vicinity – in part because of the environmental nuisance but often

[65] RHO, *ARMC (Bombay)*, *1905–06*, 196.
[66] One index of this was that in England and Wales in 1911–14 the infant mortality rate stood at 172 per 1,000: in Bombay in 1918–22 it averaged 572 per 1,000 live births and peaked in 1921 at 667 per 1,000. A. R. Burnett-Hurst, *Labour and Housing in Bombay: A Study in the Economic Conditions of the Wage-Earning Classes in Bombay* (London: P. S. King & Son, 1925), 35.
[67] Jones, *Manual*, 45–46.   [68] Bombay, Public, nos 22–26, 10 February 1834, IOR.
[69] Hehir, *Hygiene*, 398.

for religious or caste reasons as well. For instance, in 1887, residents complained that dust from a bone-crushing plant near Calcutta was falling on a sacred tank and polluting it – a claim the authorities denied.[70] In reality, those who were most at risk from the pollution and poison were often the workers themselves, the poor, low-caste men, women and children who toiled in lime kilns, tanneries and dyeing works with little, if any, protection against contact with caustic substances or the inhalation of toxic fumes. One 'offensive' trade in Bombay was the recovery of gold and silver thread from embroidery and old clothes by men called *jarriwalas*. The process of extraction, in enclosed, unventilated premises, produced 'dense clouds of irritating gases' from charcoal braziers.[71] The choking fumes were an example of the many airborne pollutants to which the urban poor were exposed, and yet, when the stench and smoke drifted into more affluent neighbourhoods, the *jarriwalas* were seen as culprits rather than victims. A further example of how the occupations of the poor were seen to impact negatively on the urban environment was rice milling. By the early twentieth century, power-driven rice mills had become common in city lanes and backstreets, producing husked and crudely polished rice for working-class consumers. To economize on coal and oil, rice husks, straw and other waste were used to fuel the engines, creating dense, choking smoke. In heavily populated wards of Calcutta, the smoke, smuts and sparks provoked complaints from residents and worried the authorities. However, it was recognized that coal was prohibitively expensive for such small-scale operations and that urban rice mills (much like milch cows in Bombay) were too important to the subsistence of the poor to be suppressed or expelled.[72] Nor, for much the same reasons, was it thought practicable to curb the acrid smoke that wafted across the city from countless cow dung fires, for this was 'the only fuel available for the poorer classes'.[73]

Ramachandra Guha and Juan Martinez-Alier have argued for what they term 'the environmentalism of the poor'.[74] Their discussion principally relates to present-day activism among the poor of Asia, Africa and Latin America in response to the threatened destruction of their environment and slender subsistence. This indigenous, need-based ideology and praxis is contrasted to the environmentalism of middle-class Westerners

[70] *Annual Report of the Sanitary Commissioner for Bengal, 1887*, 51.
[71] Turner, *Sanitation*, 782; Lemuel Lucas Joshi, 'Report of the Analytical and Bacteriological Work Done in 1914', in *ARMC (Bombay), 1914–15*, 119.
[72] ARBSNC, 1920, 2–3; ARBSNC, 1921, 3; ARBSNC, 1922, 3.
[73] John Beames, Commissioner, Burdwan Division, to Secretary, Bengal, Judicial, 9 June 1882, Bengal, Judicial, nos 192–19, February 1883, IOR.
[74] Ramachandra Guha and Juan Martinez-Alier, *Varieties of Environmentalism: Essays North and South* (London: Verso, 1997), ch. 1.

whose agenda is not primarily driven by local subsistence concerns but by wider ecological, moral and aesthetic arguments. The 'environmentalism of the poor' is a concept that can usefully be applied to colonial India, especially to popular opposition to state forest regulations from the 1870s onwards. But the term fits better with rural resistance than with big-city governance. Urban activism over the kinds of issues identified in this chapter, from contaminated water to smoke pollution and 'offensive trades', either arose from government agencies and municipal health officers or came from the higher castes and middle classes seeking to protect their neighbourhoods or defend their social rights and religious sensibilities. This anticipates what has become known in India as 'bourgeois environmentalism', in which urban pollution and environmental degradation are seen as coming, in no small measure, from the occupations and habitations of the poor. This hostile attitude often surfaced in relation to the cities of British India. Thus, according to Bombay's Health Officer in 1866, the bulk of the city's garbage came from 'amidst the dwellings of the poorest inhabitants, who are naturally the filthiest in their habits and customs, and are the great filth producers in the city'.[75] The sanitary commission in Calcutta in 1885 likewise referred to the city's bustees (slums) as being 'long recognized as a great, if not the greatest, blot on the sanitary conditions of Calcutta'.[76] The poor represented the unmodern, the atavistic, at the heart of India's cities, and their responsibility for the unsanitary state of the city often served to detract from the irresponsibility or the culpability of others.

The history of colonial Bombay is replete with instances where influential social groups or powerful property owners, European and Indian, protested through the press, by means of letters and petitions and, most effectively, through the courts against offending sources of pollution. This activism was even directed against the municipal authorities themselves. In 1870, the European members of Bombay's Byculla Club protested against the 'dense clouds' of acrid smoke produced by the burning of street sweepings and municipal waste on the adjacent Tardeo Flats.[77] In 1872, the municipality was threatened with legal action over the dumping of night soil in Bombay harbour, and in the same year a resident won an injunction against the discharge of raw sewage into the sea from the inaptly named Love Grove pumping station, which had begun operating only three years earlier.[78] In 1873, a recently opened night soil depot on Jail Road had to be closed after a successful suit against the municipality

---

[75] RHO, *ARMC (Bombay), 1866*, 4.   [76] *Report of the Commission*, 1.
[77] RHO, *ARMC (Bombay), 1870*, 2.   [78] RHO, *ARMC (Bombay), 1872*, 10.

by the owner of a nearby house.[79] In 1902, construction of a public latrine on Bohra Street in Colaba was only able to proceed after the defeat of a case brought by a local resident.[80] Unlike more recent 'public interest' litigation in India, these earlier legal cases, while they might invoke a sense of public well-being, were primarily mounted in defence of individual property interests or to protect the sanctity and salubriousness of neighbouring temples and other religious sites. The Calcutta and Bombay smoke commissions (to which we will turn shortly) received numerous complaints from the public, most of which appear to have come from indignant middle-class residents. In Calcutta, these averaged about fifty a year during the 1920s and 1930s, and most were thought sufficiently justified for the municipality to act upon.[81] Here, too, is evidence of the emergence of an Indian public sphere in which pollution, poisoning and environmental governance were issues of wider, and not just scientific, concern.[82]

Industrial cities, urban poisons

By the late nineteenth century, atmospheric pollution was already a long-standing issue in urban India, but it acquired greater intensity once burgeoning industries – led by jute in Calcutta and cotton in Bombay – began to fill air, water and soil with new sources of contamination. Among the most visible of these pollutants was the dense smoke belching from factories and workshops, railway yards, munitions factories, printing works and steamships docked in Bombay harbour or moored along the Hooghly. The atmospheric pollution of this kind dwarfed in scale and toxicity that of earlier 'offensive trades'. But the new industries were a consequence of the rise of local and international capital – Indian and European enterprises that commanded far more political clout and economic influence than tallow makers and butchers had previously done, and, once established in the city, were far more difficult either to regulate or relocate elsewhere. As with the Poisons Act of 1904, opposition to greater control came not

[79] RHO, *ARMC (Bombay), 1873*, 72.
[80] RHO, *ARMC (Bombay), 1902*, 120–21. On the unpopularity of latrines, see Mann, 'Delhi's Belly', 24.
[81] ARBSNC, 1919–20, 3.
[82] Continuing public concern about smoke and river pollution in urban India is clear from press reports: e.g., *ToI*, 4 December 1925, 15; ibid., 20 October 1927, 14; ibid., 23 May 1929, 13; ibid., 17 August 1933, 15. For pollution and 'the constitution of a public', see Amita Baviskar, Subir Sinha and Kavita Philip, 'Rethinking Indian Environmentalism: Industrial Pollution in Delhi and Fisheries in Kerala', in Joanne Bauer (ed.), *Forging Environmentalism: Justice, Livelihood, and Contested Environments* (Armonk, NY: M. E. Sharpe, 2006), 212.

Figure 8. Steamers and smoke on the River Hooghly, Calcutta, c. 1890.

from the public at large, and not from the poor, but from commercial and industrial interests like the jute mill owners in Calcutta.[83] To give one example, in 1915, business organizations in the city opposed an amendment to the Bengal Smoke Nuisances Abatement Act which would prohibit ocean-going steamers on the Hooghly from firing up their boilers for two hours before their departure, thereby generating dense clouds of acrid smoke that drifted across the city centre (see Figure 8). The complaint was that this was a vexatious constraint on commercial freedom and was bound to have adverse consequences for the operations of the steamship companies. But, in this instance, protests about pollution, led by Indian members of the provincial legislature, prevailed over the views of the predominately European business lobby, and the amendment was passed.[84]

Smoke, as one of the most visible forms of pollution, was a crucial issue in urban environmental governance. Michael Anderson has described attempts in Calcutta to curb the 'smoke nuisance', and there is no need to repeat here material he so ably presented.[85] Suffice it to say that

[83] C. W. Bolton, Chief Secretary, Bengal, to Secretary, Bengal Chamber of Commerce, 2 March 1900, Bengal, Judicial, no. 5, March 1900, IOR.

[84] L/P&J/6/1374: 2315, IOR.

[85] Michael R. Anderson, 'The Conquest of Smoke: Legislation and Pollution in Colonial Calcutta', in David Arnold and Ramachandra Guha (eds), *Nature, Culture, Imperialism: Essays on the Environmental History of South Asia* (Delhi: Oxford University Press, 1995), 293–335.

atmospheric pollution was no less troublesome an issue in Bombay by the 1900s, where a pall of dark factory smoke hung over the city between November and March.[86] As Anderson noted, air pollution was principally judged by visual criteria, by the appearance of the smoke from factory chimneys, by its relative darkness, density and general 'nuisance' quality rather than by means of chemical analysis or in terms of any discernible effect on human health. A series of enquiries were conducted, of which the most influential was the report made in 1903 by a British expert, Frederick Grover, on smoke pollution in Calcutta. In this and many other instances, the experience of Britain's industry was cited and metropolitan expertise employed to resolve the kindred problems of Indian cities.[87] But claims were also made about the peculiar physical and social circumstances of India. Indian coal was said to be of poorer quality than British coal and so produced a greater quantity of dense smoke; Indian stokers were less competent than their British counterparts and so less able to control the volume and density of smoke generated by their boilers.[88] 'It is not to be expected', Bengal's Chief Engineer remarked in 1891, 'that with country coal and the average native fireman we will ever reach the state of perfection of smoke abatement that exists in the London and Liverpool districts'. Even so, he added, 'a great improvement could be made without seriously interfering with any interest'.[89] The determination to improve air quality *without* unduly disturbing vested interests was a significant admission of moderation, if not weakness.

What appeared on paper to be a wide-ranging system of inspections, warnings and fines was instituted in both Bombay and Calcutta, beginning with the Bengal Smoke Nuisances Act of 1905. Smoke abatement legislation represented a new departure in the prolonged attempt to regulate urban environments.[90] It formed part of a wider programme of environmental management introduced across British India from the 1870s onwards, with growing intensity by the 1890s and 1900s, which encompassed forests, fisheries, wildlife and waterways as well as cities.[91] But there were practical limitations to what such measures could

[86] RHO, *ARMC (Bombay), 1907–08*, 227.
[87] Frederick Grover, *Report on the Abatement of Smoke Nuisance in Calcutta* (Simla: Government Central Printing Office, 1903); also, William Nicholson, *Report on Smoke Nuisances and Their Abatement in Calcutta* (Calcutta: Bengal Secretariat Book Depot, 1906). On the British experience, see *Report of the Smoke Abatement Committee, 1882* (London: Smith, Elder, 1883).
[88] Grover, *Report*, 5–12.
[89] R. Rushby, 'Abatement of Smoke Nuisance', Bengal, Judicial, nos 28–29, January 1892, IOR.
[90] For the Bengal act, see Nicholson's *Report*.   [91] Alley, *On the Banks*, 134–39.

realistically achieve. An earlier piece of legislation, the 1863 Smoke Abatement Act in Bengal, rather like the Bombay Poisons Act of 1866, was said by 1903 to have 'entirely failed', having resulted in only eighteen successful prosecutions in forty years.[92] New legislation in the 1900s proved only partly effective. The reports of the smoke commissioners appointed to oversee implementation of the abatement measures detail the number of sightings made of factory smoke, the number of inspections held and warnings issued. For instance, in 1919–20, Bengal's inspectors made 9,943 smoke observations, conducted 3,288 factory visits and issued 16 warnings to offending factories and their owners throughout the province.[93] Overall, the commissioners expressed some optimism about the improvement in urban air quality, and some progress was undoubtedly made, including a shift from heavily polluting coal to relatively smokeless coke for both domestic fuel and factory boilers. Yet reference was also made in the reports to failures and lapses that made the commissions' achievements appear less impressive. Calcutta's first smoke inspector was removed from office because he was not sufficiently energetic in pursuing prosecutions.[94] The smoke inspectorate had only a small staff at its disposal, and this was further diminished by European manpower shortages during the First World War. There was frustration that more could not be done to punish persistent offenders or to levy fines large enough to act as a real deterrent. 'All is done with kindness', remarked the President of Bombay's Smoke-Nuisances Commission in 1915, implying that in the face of factory owners' evasion and resistance, persuasion and negotiation were often the only possible options.[95] Matters remained largely unchanged in the interwar years, but with the outbreak of the Second World War, many restrictions on smoke emissions were relaxed. The government was anxious to meet mounting wartime demand for cotton, jute and other industrial goods, and pollution controls which restricted production were accordingly lifted. This was meant to be a temporary measure, but such concessions, once made, were not easily rescinded.[96]

Factory smoke was the most visible manifestation of urban pollution, but it was not necessarily the most damaging or pervasive. Although the mechanization of transport, combined with a growing body of provincial laws and municipal regulations, helped reduce animal waste and

---

[92] Grover, *Report*, 5.   [93] ARBSNC, 1919–20, appendix.

[94] Bengal, Judicial, nos 72–73, May 1902, IOR; Bengal, Judicial, nos 1–6, July 1902, IOR; Bengal, Judicial, nos. 10–11, August 1902, IOR.

[95] *Bombay Smoke-Nuisances Commission, 1914–15*, 2; RHO, *ARMC (Bombay)*, 1908–09, 28.

[96] ARBSNC, 1944, 7.

moderate smoke emissions, other types of pollution began to emerge instead. Traffic pollution caused by car and lorry exhaust was first considered as a potential risk to health in Bombay in 1906, but this was quickly dismissed on the grounds that it was rapidly dissipated by sea breezes blowing across the island.[97] It was not until the 1930s that cars, trucks and buses came to be seen as a more serious danger to health and one which could by then be more precisely monitored and quantified.[98]

Moreover, in the city – at home, in the street, in factories and workshops – toxicity took on more sinister forms, posing new hazards to human health and well-being. Pollution, adulteration and contamination could be more than a 'nuisance': they could incapacitate and kill. By the 1930s, fresh lines of enquiry were being pursued in relation to urban toxicity. As Bengal's Chemical Examiner, K. N. Bagchi conducted research on arsenic and lead in Calcutta. One investigation involved examining arsenic levels in common items of Indian diet, a revisiting of the old, but never absent, issue of contaminated or poisoned food.[99] Although the tests showed a generally low arsenic content, the studies were of forensic value, for without knowing what constituted normal levels of arsenic in the human body, it was hard to determine the deliberate use of arsenic in murder or suicide.[100] Other findings from the study were, however, less reassuring, especially in relation to lead poisoning. Lead had been mentioned occasionally as an accidental hazard in India, as a consequence, for instance, of the use of lead oxide to treat syphilis. But its toxicity was not given much prominence until studies in Europe and North America began to reveal the danger lead poisoning might pose. Bagchi asked whether a similar risk existed in India where industrialization – and so the industrial and domestic use of lead – was less developed: as with river pollution, there was a presumption that India remained in a largely pre-industrial age to which issues of industrial pollution and poisoning bore little relevance. Overall, the incidence of lead in the sample population was found to be substantially lower than in the West, though Europeans and Anglo-Indians showed relatively high, if seldom clinical, levels of lead.[101] One anomaly, however, concerned married Hindu women, whose hair and skin samples revealed high

[97] RHO, *ARMC (Bombay), 1906–07*, 207–08.
[98] *Bombay Smoke-Nuisances Commission, 1934–35*, 7–8.
[99] K. N. Bagchi and H. D. Ganguly, 'Arsenic in Food', *IMG* 76 (1941): 720–22.
[100] K. N. Bagchi and H. D. Ganguly, 'Arsenic in Human Tissues and Excreta', *IMG* 72 (1937): 477–81; A. C. Bose, 'A Modified Method of Estimating Arsenic-Content of Indian Food-Stuffs', *IJMR* 22 (1935): 697–700.
[101] K. N. Bagchi and H. D. Ganguly, 'Lead in Urine and Faeces', *IJMR* 25 (1937–38): 147–54; K. N. Bagchi, H. D. Ganguly and J. N. Sirdar, 'Lead in Human Tissues', *IJMR* 26 (1938–39): 935–46.

concentrations of lead. This was attributed to the substitution of the *sindur* (vermillion) Hindu wives wore on their hair parting with cheaper red lead, which caused lead to accumulate in the hair and scalp at levels more than three times higher than those found among Muslim women, who never used *sindur*.[102]

No less significant than this evidence of domestic lead poisoning was its industrial occurrence. *McNally's Sanitary Handbook* cited the danger posed by industries that used lead, antimony, arsenic and phosphorus – all of which 'produce chemically poisonous particles of dust, and frequently give rise to diseases amongst those who work with these substances'.[103] In 1921, J. J. Campos published an article on plumbism among Calcutta print workers. He began by noting that, while much attention had recently been paid to industrial health in Europe and North America, the subject was 'almost completely neglected' in India. Despite the presence in Calcutta of mills and factories employing thousands of workers, 'scarcely any organized study is made ... and no morbidity statistics [are] collected'.[104] Pollution, still more poison, was a barely recognized aspect of what happened in the workplace and to the health of industrial workers. Lead was extensively used in paint, in making shot, in the manufacture of cooking pots and in the type used in printing works. Compositors repeatedly handled lead type, even holding it in their mouths as they worked, and they and other print workers constantly inhaled air thick with fine lead particles. The symptoms of lead poisoning were not hard to detect – a blue line appeared on the gums, and sufferers experienced anaemia, constipation, colic and eventual paralysis. But, Campos found, as soon as print workers in Calcutta became ill, they tended to leave work and quit the city. 'Probably they pass to the hands of a village doctor or a Kaviraj for treatment, while the causal relation between his [*sic*] former occupation and the disease is forgotten.'[105]

Campos' article was praised in the medical press at the time, but it was years before follow-up studies were made.[106] It then transpired that while print workers were particularly at risk, they were not alone. Further evidence of the problem came to light through an investigation of labour

---

[102] K. N. Bagchi, H. D. Ganguly and J. N. Sirdar, 'Lead Content of Human Hair', *IJMR* 27 (1939–40): 777–91; K. N. Bagchi, 'Incidence of Lead Poisoning among Hindu Women and Children', *IMG* 76 (1941): 23–29.

[103] Russell (ed.), *McNally's*, 81.

[104] J. J. Campos, 'Chronic Lead Poisoning in the Printing Presses of Calcutta', *IMG* 56 (1921): 175–76.

[105] Ibid., 177.

[106] Editorial, 'Industrial Medicine and Hygiene in Bengal', *IMG* 57 (1921): 182; R. H. Candy, 'A Note on the Prevalence of Lead Poisoning in India', *IMG* 68 (1933): 136–37.

conditions in the chemical industry in Bengal in 1946. This showed that 'gas attacks' were not infrequent among workers handling white and red lead. Gloves and respirators were sometimes supplied to the workers but, from their reluctance to wear them, seldom used. Workers who fell ill after months of breathing an atmosphere thick with lead dust simply disappeared from the workforce, their fate unknown. Factory inspections were few and far between, with some chemical works in the Calcutta area not inspected for three years prior to the study.[107] By the early 1950s, occupational lead poisoning had become a hazard in the increasing number of factories – an estimated fifty in Calcutta alone – that used lead in their operations. Lead poisoning could easily be determined by scientific tests, but the collection of data in the workplace was not always possible. Attempts to study plumbism in the factory were hampered by 'the popular belief in managements as well as responsible medical administrators that there is no lead hazard in Indian industries'. Managers objected to having their employees examined while the workers themselves were often wary, fearing that they would lose their jobs if lead poisoning were detected. Assessing the risk of lead poisoning called for more than tests on individual workers: it required careful analysis of the air in factories and workshops, something which could only be done with the agreement of managers and foremen. Officially, responsibility for monitoring lead pollution rested with the factory inspectorate, but inspectors were few in number and had so many other duties to perform during their occasional visits that there was little time, or perhaps inclination, to conduct technically demanding tests. As a result, the number of cases of poisoning in Indian factories from lead and other metals was probably much higher than official statistics suggested.[108] A further study in 1952 concluded that in India, unlike the United States and Britain, the scientific study of lead and other workplace poisons, and the need for action to redress the problem, was still only 'just on the threshold of consciousness'.[109] As India entered the era of its independence, the systematic investigation and monitoring of industrial poisoning and pollution was still a long way off.

More perhaps than the long-standing but highly visible issue of 'smoke nuisance', lead pollution in the home and factory highlighted the problem of urban toxicity and the difficulty of its detection and control. Here was a

[107] B. P. Adarkar, *Report on Labour Conditions in the Chemical Industry* (Simla: Manager of Publications, Government of India Press, 1946), 45.

[108] M. K. Chakraborty, M. N. Rao and B. Banerji, 'A Study of the Occupational Lead Hazard in Selected Indian Industries', *IJMR* 38 (1950): 429–56.

[109] C. V. Sabnis, 'Evaluation of Lead Hazard in a Pigment Manufacturing Concern', *IJMR* 40 (1952): 53. For the continuing danger of lead pollution from Calcutta factories, see *ToI*, 6 April 1992, 8.

source of poisoning over which the state and public health agencies had little effective control. But this was more than a question of defective or insufficient governance: it was also the consequence of a culture of denial or collective unconcern. Factory owners and workshop mangers were loath to allow scientific studies to be conducted among their workforce, fearing that they would prove detrimental to their command of labour and their profit making. Trade union organizers, although concerned about other aspects of worker welfare, seemed unaware of, or indifferent to, the toxic dangers workers faced. And sometimes workers themselves failed to realize the dangers involved in handling toxic substances and the need to protect themselves.

Questions of vulnerability and responsibility bring us back to the idea, put forward in earlier chapters, of poison tales. In this instance, we can see industrial poisoning as involving a series of overlapping narratives, each bearing a different meaning. In April 1942, in wartime Bombay, one worker died and another fell seriously ill shortly after they had unloaded a consignment of a bright yellow powder, the chemical paranitroaniline used as a dye in the textile industry. One of the packages had begun to leak, and the workers amused themselves by throwing handfuls of it at each other and rubbing it over their bodies.[110] It is clear from the behaviour of the two men that the story they told themselves, as they threw the yellow powder around and smeared it on their arms and faces, was that this was just like Holi, the Hindu spring festival. This cultural reading of the episode was not one given at the time, though it must have been obvious to contemporaries. But it resonates with a second story: in March 1978, fifty children were hospitalized after exposure to a coloured powder containing paranitroaniline during the celebration of that year's Holi. This caused a poison scare in the city and alarm lest other colours used in the festival might also be toxic.[111] However, the version of the poison story given in 1942 was that presented by Bombay's Assistant Chemical Analyser. He ignored Holi, concentrating instead on the virtual absence from Indian medical textbooks of any discussion of industrial chemicals and the hazards they posed. As industrialization gained pace, he warned, medical education and toxicological science would have to keep up and address new toxic dangers.[112]

Poison and the post-colony

If India's toxic history did not end with the passing of the Poisons Act of 1904, neither did it conclude with independence and partition in 1947.

[110] R. Lobo-Mendonca, 'A Note on Paranitroaniline Poisoning', *IMG* 77 (1942): 673–75.
[111] *ToI*, 27 March 1978, 1.   [112] Lobo-Mendonca, 'Note', 675.

Indeed, poison as personal tragedy, environmental hazard and collective calamity has loomed large in many accounts of India and its South Asian neighbours since the 1940s. From the Bhopal disaster of 1984 to farmer suicides with toxic pesticides and arsenicosis in the Bengal Delta, poison seems almost to define India's post-colonial condition.[113] With what justice? How far can this recurrent trope of toxicity be seen as a further restatement of Orientalist negativity and a belittling of Indian achievement? Or, decades after independence, as a projection of the ills and errors of the colonial era into a very different age of national sovereignty and global interdependence?

There are many ways of thinking about poison, pollution and the post-colony. One is to emphasize the colonial inheritance and to argue that the British left a baleful legacy – of insufficient governance and mismanaged opportunities – that India still lives with or has only gradually been able to redress. Or, more positively, to suggest that at least some of the laws and agencies established by the British – including the chemical examiners – continued to have a regulatory role well into the post-colonial era. But we should also note the changes that independence itself brought about, especially the technological optimism (but also complacency) that surrounded that momentous event and the developmental imperatives it helped unleash. One part of this transition from colonial rule to post-colonial regime was a tendency to sacrifice the controls over urban, industrial and environmental toxicity built up, however partially and ineffectively, during the British era. As we have already seen in this chapter, all too little was done after 1947, and even earlier from the Second World War onwards, to keep up with the accelerating pace of industrial growth and urban expansion. Although laws were passed to protect employees in dangerous trades, many workers failed to take advantage of such laws or were prevented by their employers from effectively doing so.[114] While scientific investigations into industrial poisoning and environmental pollution in India after 1945 closely followed British and, increasingly, American research questions and methodologies, they were frequently hampered by a lack of funding and expertise and by inadequate laboratory facilities and equipment. And while some studies, like those already referred to, relating to air pollution and lead poisoning in factories detailed the dangers posed, yet others seemed remarkably

---

[113] Upamanyu Pablo Mukherjee, *Postcolonial Environments: Nature, Culture and the Contemporary Indian Novel in English* (Basingstoke: Palgrave Macmillan, 2010); Rob Nixon, *Slow Violence and the Environmentalism of the Poor* (Cambridge: Harvard University Press, 2011).

[114] For India's labour laws, see Emmanuel Teitelbaum, 'Was the Indian Labor Movement Ever Co-Opted?', *Critical Asian Studies* 38 (2006): 409.

complacent and argued that conditions in factories or in relation to rivers and urban water supplies gave no great cause for alarm.

For instance, an inquiry by researchers at Calcutta's All-India Institute of Hygiene and Public Health in 1950 examined the sanitary condition of the Hooghly, using guidelines laid down by the US public health services, but the investigation was 'circumscribed by the very limited resources and facilities available at our disposal'. They listed and tabulated the see-mingly vast quantities of effluent and liquid waste discharged into the river by sewage outfalls, jute mills, paper mills, tanning works and che-mical and pharmaceutical factories. Remarkably, however, they con-cluded that, given the effects of dilution by river flow and tidal reach, there was little sign of serious pollution: 'the load of pollution from industrial wastes and the sewage is not considerable'.[115] An investigation into Calcutta's jute mills published in 1951 similarly noted that more might – and should – be done to improve ventilation and reduce air temperatures, and so make working conditions more bearable, but other-wise 'the environmental conditions in jute mills appear to be fairly satisfactory'.[116] Perhaps this apparent complacency resulted from too great a willingness to accept the managerial view, or by a readiness to see India's rivers as immune to pollution and capable of negating all the poisons and pollutants discharged into them. But possibly, too, in the months and years that followed independence, there was an inclination, even within the Indian scientific community, to believe that the science of the colonial era had been too negative and constraining. The function of science in the post-colony was to serve the nation, combat disease, promote growth and foster social well-being, not to carp and cavil over the seemingly minor side effects of industrial development and economic autarchy.

Poison has a long connection with India's development history. In the nineteenth century, the quest for 'improvement' took many forms – the introduction of new crops or superior varieties of existing ones, the commercialization and mechanization of agriculture, the spread of irriga-tion, the cultivation of wasteland, the harnessing of water and forest resources. But it was the improving task, too, of sanitarians and scientists, to overcome the 'poison' of miasmatic swamps and jungle vegetation or urban slums and graveyards, and so to enable Indians to have healthier, and hence more economically productive, lives. As the understanding of

---

[115] G. K. Seth and T. R. Bhaskaran, 'Effect of Industrial Wastes Disposal on the Sanitary Condition of the Hooghly River in and around Calcutta', *IJMR* 38 (1950): 346, 353; T. R. Bhaskaran, 'A Plea for Water Pollution Research', *IMG* 82 (1947): 750–52.
[116] K. Subrahmanyam and N. Majumdar, 'Environmental Conditions within Jute Mills', *IJMR* 39 (1951): 623.

disease aetiology became more refined in the 1890s and 1900s, so poison came to represent not just a manifestation of diseases to be overcome but also the means of achieving that goal by poisoning rats or using antimonials to treat kala-azar. As India began by the 1930s to edge away from a colonial notion of improvement to a nation-building agenda, so poison in its more positive guise began to present other developmental possibilities.[117] One of these was the poisoning of the insects that devastated crops or spread animal as well as human diseases. There was an Orientalist view that insecticides were unacceptable to Indian sensibilities, especially to Hindus and Jains who objected to the taking of any life, even that of insects. In 1908, the government entomologist H. Maxwell-Lefroy observed: 'The Eastern mind has not yet fully grasped the idea that insects could be or should be killed by hand-picking, far less by such a method as poisoning the plant with arsenic.'[118] Before 1900, colonial initiatives to thwart famine and augment agricultural productivity had focused on increasing the supply of water for irrigation or the promotion of new or improved crop varieties. But the increasingly scientific tenor of agricultural research in India fostered a belated interest in the more systematic application of chemistry to agriculture and the possibility of using chemical fertilizers to enhance yields and insecticides to combat insect predation. Pesticide use rose only gradually, however, and mainly in relation to what had become India's greatest killer and leading obstacle to development – malaria. Experiments were conducted in the 1930s in spraying Paris green (another arsenic compound) on stagnant or slow-moving water to kill breeding mosquitoes and their larvae.[119] Hopes then turned to pyrethrum as a plant poison that could be used to protect homes and workplaces by destroying malarial mosquitoes. But the most dramatic advance in the battle against malaria came at the end of the Second World War with DDT (dichlorodiphenyltrichloroethane).

DDT has been hailed as a 'revolutionary factor' in India's post-war public health and development programmes.[120] It decimated insect populations, almost eliminated human and animal diseases spread by

[117] See David Ludden, 'India's Development Regime', in Nicholas B. Dirks (ed.), *Colonialism and Culture* (Ann Arbor: University of Michigan Press, 1992), 247–87; Benjamin Zachariah, *Developing India: An Intellectual and Social History, c. 1930–1950* (New Delhi: Oxford University Press, 2005).

[118] H. Maxwell-Lefroy, *Indian Insect Pests* (Calcutta: Office of the Superintendent of Government Printing, India, 1906), 75–76.

[119] As in the Agency hill tracts of the Madras Presidency: GO 2268, Madras, Local Self-Government (Public Health), 16 September 1930, TNA; GO 407, Madras, Local Self-Government (Public Health), 26 February 1932, TNA.

[120] B. H. Farmer, *Agricultural Colonization in India since Independence* (London: Oxford University Press, 1974), 45.

flies, fleas, lice, ticks and mosquitoes, reduced malaria deaths from over 800,000 a year on the eve of independence to almost zero by 1965 and enabled agricultural expansion into regions like the Terai on the India-Nepal border that endemic malaria had previously rendered almost uninhabitable.[121] But it was not just the extraordinary potency of DDT as an insecticide that made it so appealing to public health experts in India. It was also the timing of its appearance. The first extensive trials with DDT came just as British rule was ending and independent India coming into being. At the very moment when India needed land for demobilized soldiers and refugees rendered homeless and landless by partition, so DDT made new tracts of malaria-free land suddenly, almost magically, available. At the precise juncture when India's government was looking to shed the seeming lethargy and indifference of the colonial era and present a new national purposefulness in science and self-sufficiency, DDT arrived to show how an energetic state, wedded to development and allied to modern science and technology, could overcome problems of health, productivity and even population growth that had appeared intractable in the past. Like the killing of rats and fleas to curb plague in the 1900s, so DDT in the 1940s and 1950s presented a means of bypassing the social and cultural issues that had hitherto obstructed interventionist public health by dealing directly with the insect vectors and parasites that caused or transmitted disease. It was noted as a triumph of DDT trials in south India that house spraying met with no objections from purdah women and that Jains did not protest against this wholesale culling of insect life. Even Gandhi gave his blessing.[122] DDT was seen as having immense value in combating precisely the kinds of tropical diseases that had for so long delayed India's development. India now entered the 'DDT era' of public health.[123]

Even at the time of its introduction, there were somewhat divergent views as to DDT's toxicity. Some scientists in Europe and America expressed a degree of caution about its use, warning that although DDT combined 'high and general insecticidal power with low toxicity to mammals', it could endanger human health if accidentally applied to the skin or swallowed in large doses.[124] Indian researchers tended to set such caveats aside: DDT could be used with absolute confidence. Even those

[121] R. S. Srivastava, 'Malaria Control Measures in the Tarai Area under the Tarai Colonization Scheme', *IJM* 4 (1950): 151–65.
[122] D. K. Viswanathan, *The Conquest of Malaria in India: An Indo-American Co-Operative Effort* (Madras: Company Law Institute Press, 1958), ch. 6.
[123] Ibid., ch. 4.
[124] P. A. Buxton, 'The Use of the New Insecticide DDT in Relation to the Problem of Tropical Medicine', *Transactions of the Royal Society of Tropical Medicine and Hygiene* 38 (1945): 391.

who did the spraying, and might experience slight irritation to the nose and mouth, were assured that their health would not be affected adversely.[125] In the United States and in many other parts of the world, scientific endorsement for DDT crumbled in 1962 when, fifteen years into India's independence, Rachel Carson launched what one critic called her 'poetic vilification of DDT'.[126] As another commentator put it, Carson used pesticides as a symbol to communicate her view of the dire consequences of humans' misguided attempts to master nature. After *Silent Spring*, 'The public felt betrayed, and science and technology, previously considered valuable allies, were seen as nature's enemies'.[127]

Following the lead of the United States in 1972, governments around the world prohibited the use of DDT and other insecticides with a detrimental effect on humans and wildlife. It has been proposed that there should be international agreement to phase out DDT entirely by 2020. But the Indian government has refused to follow suit, and India remains one of the few countries that still manufactures DDT and approves its insecticidal use.[128] The argument is that DDT is safe, that there has not been a single death from DDT poisoning and that it remains a necessary check against resurgent malaria.[129] The revitalization of Indian agriculture through the Green Revolution of the late 1960s and early 1970s made it all the more imperative that Indian farmers have access to large quantities of chemical fertilizers and pesticides.[130] By the mid-1970s, India had become the fifth largest pesticide market in the world, and, in order to exploit this demand, chemical factories were established in India. Among these was the Union Carbide plant in Bhopal in 1969, where highly toxic methyl isocyanate was manufactured for use in commercial pesticides.[131]

[125] I. M. Puri and Rajindar Pal, 'Studies of Some Insecticides against *Anopheles* Adults and Larvae', *IJM* 1 (1947): 133–58; S. L. Kalra, V. P. Jacob and K. N. A Rao, 'Effect of DDT and BHC on *Ornithodoros* Ticks', *IJMR* 38 (1950): 457–66.

[126] Donald R. Roberts and Richard Tren, 'Did Rachel Carson Understand the Importance of DDT in Global Public Health Programs?', in Roger Meiners, Pierre Desrochers and Andrew Morriss (eds), Silent Spring *at 50: The False Crisis of Rachel Carson* (Washington: Cato Institute, 2012), 167.

[127] C. F. Wilkinson, 'The Science and Politics of Pesticides', in Gino J. Marco, Robert M. Hollingworth and William Durham (eds), Silent Spring *Revisited* (Washington: American Chemical Society, 1987), 27.

[128] 'India Opposes 2020 Deadline for DDT Ban', www.downtoearth.org.in/content/india-opposes-2020-deadline-ddt-ban.

[129] K. N. Mehrotra, 'Use of DDT and Its Environmental Effects in India', *Proceedings of the Indian National Science Academy* B51 (1985): 169–84.

[130] On US support for India's anti-malaria campaign and use of DDT, see David Kinkela, *DDT and the American Century: Global Health, Environmental Politics, and the Pesticide That Changed the World* (Chapel Hill: University of North Carolina Press, 2011), 101–02.

[131] Larry Everest, *Behind the Poison Cloud: Union Carbide's Bhopal Massacre* (Chicago: Banner Press, 1986), 45–51; Radhika Ramasubban, 'Profit Against Safety', *Economic and Political Weekly* 19, nos 51–52 (22–29 December 1984): 2147–50.

The poison gas leak at Bhopal in December 1984 was not just 'the most massive industrial disaster in history', but also an immense human catastrophe that has left a lasting legacy of suffering, denial and recrimination.[132] In terms of human lives lost and the extent of the suffering caused, Bhopal dwarfed any of the poison episodes of the colonial era, and there was much about the episode that was specific to the period and the place in which it happened. But it also re-enacted, on a far greater and more violent scale, many of the themes – of insufficiently regulated industrial and urban toxicity, of management negligence and profit seeking above safety, of poor governance and even poverty – that had been evident in India's toxic history for decades.[133] Like the colonial regime before it, the Government of India had passed laws for environmental protection – including acts to control water and air pollution in 1974 and 1981 – but, given the size of the problem, that legislation remained largely ineffective and pollution continued relatively unchecked. Controls over poison and pollution in the workplace had been tightened up and extended since the 1940s, but still failed to protect workers and exposed those who lived in adjacent slums and inner-city shanties to ever-present dangers.[134] Tragically for Bhopal, too little was done to ensure the safety of either those who worked within the plant (where there had been several accidents and leaks between 1978 and 1984) or those who lived in close proximity to it. Development, seen as a solution for the problems of poverty and backwardness that a negligent or self-interested colonialism had either created or exacerbated, called for pesticides to be manufactured on an industrial scale, and an American enterprise and its Indian subsidiary helped supply that demand. But in this instance, and many like it, toxicity, once embraced, could not so easily be mastered.

The agricultural use of hazardous chemicals carried many of the dangers of toxicity evident for decades in industrial cities like Bombay and Calcutta into the countryside, with adverse consequences for human health, wildlife and the environment. As early as 1958, the Indian press reported cases of food poisoning due to leaked insecticides and called for tighter controls on their storage, sale and disposal.[135] In India, and across the border in Pakistan, pesticide spraying of cotton and other commercial

[132] Everest, *Poison Cloud*, 78.

[133] On Bhopal, see also Ingrid Eckerman, *The Bhopal Saga: Causes and Consequences of the World's Largest Industrial Disaster* (Hyderabad: Universities Press, 2005); Suroopa Mukherjee, *Surviving Bhopal: Dancing Bodies, Written Texts, and Oral Testimonials of Women in the Wake of an Industrial Disaster* (New York: Palgrave Macmillan, 2010).

[134] A. O. Noorani, 'The Citizens' Environment', *Economic and Political Weekly* 19, nos 51–52 (22–29 December 1984): 2150.

[135] *ToI*, 14 June 1958, 7.

crops by workers inadequately protected from toxic sprays – or insufficiently aware of their dangers – has become a major health issue. It has been estimated that exposure to toxic pesticides causes at least 22,000 deaths a year in India.[136] And it is not only human lives that have been affected. In the late nineteenth century, it was cattle poisoning that attracted attention. In more recent decades, the poisoning of vultures that fed on the carcasses of cattle previously treated with the veterinary drug diclofenac led to a precipitous fall in their numbers, a decline only slowly reversed since the drug was taken off the market in 2006.[137] In a further replaying of an old theme, much as unhappy wives once crushed up their glass bangles in order to attempt suicide, so Indian farmers, trapped by rising debts and falling commodity prices, have turned, symbolically and practically, to pesticides to end their lives. By one account, more than 25,000 agriculturalists died in this way between 1997 and 2004. In parts of South Asia, suicide by self-poisoning has reached epidemic proportions.[138]

Many of the same resources and the same substances have come to figure in the toxic history of post-colonial, as of colonial, South Asia, but often in significantly different ways. In West Bengal and Bangladesh, the 1980s began to produce evidence of mass poisoning caused by arsenic – that old adversary – in groundwater brought to the surface by tube wells. Clearly, this was arsenic without the homicidal intent but a naturally occurring phenomenon aided by human agency. The installation of tube wells in rural areas had been designed to provide clean, uncontaminated water fit for human consumption and free from cholera and gastrointestinal infections that had previously caused widespread sickness and high mortality.[139] This, too, was a development scenario, initiated with laudable intentions, but with unanticipated negative results. Although since the 1940s scientific reports around the world have shown arsenic to be present in groundwater, this does not seem to have been registered as a danger with regard to South Asia until relatively recently. But by 2010 an

---

[136] Edward Broughton, 'The Bhopal Disaster and Its Aftermath: A Review', *Environmental Health* 4 (2005): 4. See also Naila Hussain, *Poisoned Lives: The Effects of Cotton Pesticides* (Lahore: Shirkat Gah, 1999).

[137] 'State of the World's Birds', www.birdlife.org/datazone/sowb/casestudy/156.

[138] Vandana Shiva and Kunwar Jalees, *Farmers Suicides in India* (New Delhi: Research Foundation for Science, Technology and Ecology, n.d.); David Gunnell and Michael Eddleston, 'Suicide by Intentional Ingestion of Pesticides', *International Journal of Epidemiology* 32 (2003): 902–09; Vendhan Gajalakshmi and Richard Peto, 'Suicide Rates in Rural Tamil Nadu, South India', *International Journal of Epidemiology* 36 (2007): 203–07.

[139] V. Govinda Raju, 'Deep Tube-Well Water of Bengal', *IJMR* 19 (1931–32): 447–55; K. Subrahmanyam, T. R. Bhaskaran and C. Chandra Sekar, 'Studies on Rural Water-Supplies', *IJMR* 36 (1948): 211–47.

estimated 36 million people in the Bengal delta region were exposed to groundwater arsenic, making this 'the largest mass poisoning of a population in history'.[140] Drinking, cooking and washing in arsenic-contaminated water causes skin lesions, a thickening of the skin on the palms and soles, black skin spots and liver enlargement, leading ultimately to ulcers, cancers and cardiovascular disease. Many of those visibly affected by arsenic poisoning, especially women, have become socially ostracized, as if suffering from that ancient stigma, leprosy. And, again, it was the poor – mostly the rural and small-town poor – who were principally affected as they were unable to turn to alternative sources of water.[141] More affluent societies are better able to adopt remedial measures to deal with arsenic contamination: poor countries like Bangladesh lack the financial and technical resources to do so.

For India, as for many other parts of the post-colonial world, poisoning has assumed a degree of menace beyond anything colonial science ever envisaged or Rachel Carson imagined in the 1960s. One response has been to see science and technology as themselves to blame, for fuelling the profit-hungry recklessness of modern industry, for the wanton neglect of nature and denial of basic human needs and for the alliance of science with a corrupt state and a rapacious but irresponsible global capitalist order.[142] An alternative reaction has been to look to science to solve the problems it has helped create – by finding new ways to control toxicity or ensure that the benefits of mobilizing poisons outweigh their ever-present dangers.[143] It would be unfair to suggest that India has not learned from the colonial past or failed to recognize the magnitude of the toxic dilemmas that decades of development have helped to unleash. Throughout the post-independence decades, there were scientists, journalists and political activists who repeatedly drew attention to the dangers of polluted rivers and poisoned farmland, of toxic pesticides and chemical spills, of

---

[140] S. H. Farooq, D. Chandrasekharam, Z. Berner, S. Norra and D. Stüben, 'Influence of Traditional Agricultural Practices on Mobilization of Arsenic from Sediments to Groundwater in Bengal Delta', *Water Research* 44 (2010): 5575; Allan H. Smith, Elena O. Lingas and Mahfuzar Rahman, 'Contamination of Drinking-Water by Arsenic in Bangladesh: A Public Health Emergency', *Bulletin of the World Health Organization* 78 (2000): 1093.

[141] Andrew A. Meharg, *Venomous Earth: How Arsenic Caused the World's Worst Mass Poisoning* (Basingstoke: Macmillan, 2005); Arsenic Policy Support Unit, *Selected Papers on the Social Aspects of Arsenic and Arsenic Mitigation in Bangladesh* (Dhaka: APSU, 2006).

[142] For example, Vandana Shiva, 'Reductionist Science as Epistemological Violence', in Ashis Nandy (ed.), *Science, Hegemony and Violence: A Requiem for Modernity* (New Delhi: Oxford University Press, 1988).

[143] Yogendra N. Srivastava, *Environmental Pollution* (New Delhi: Ashish Publishing House, 1989).

untreated sewage and proliferating urban waste.[144] The Citizens' Report on *The State of the Indian Environment* in 1982 made telling use of scientific studies and press reports to reveal the alarming extent of pollution in India, exacerbated by official neglect and apparent public indifference.[145] A second report three years later detailed the impact of the recent Bhopal disaster, but also placed in it the wider context of occupational health hazards and the toxic effects of groundwater pollution, soil erosion, dam construction and forest depletion.[146] But, in the main, the voices of these and many other critical reports have not been adequately listened to.

The Bhopal disaster helped create a more environmentally aware public in India and around the world. That catastrophe had some, if only temporary, impact on policy in India. Poison still has the capacity to shock and to scare. But, as too often in the past, a kind of inertia seems to infect the state, industries and big business appear loath to accept their social and environmental responsibilities while among the middle classes there has frequently been a tendency to blame the poor for environmental degradation and big-city 'nuisances'.[147] India's drive for global economic status and increased industrial productivity over the past twenty years has come at a high price, one that can be measured, almost literally, in terms of poison and pollution.[148] No country in the world is free from such hazards, and there is always the danger, now as in the past, of seeming to Orientalize India by highlighting its failings and deficiencies, its cultural or climatic idiosyncrasies, as though the West were innocent of toxic abuses and environmental degradation. But, in the end, as this book has tried to suggest, there is a long history to India's toxic entanglements. India's poison pasts still speak to its toxic present.

---

[144] P. P. Karan and W. A. Bladen, 'Geographical Aspects of Environmental Pollution in India', *Geoforum* 7 (1976): 51–57; Sorab J. Arceivala, 'Are We Slowly Being Poisoned?', *ToI*, 13 April 1969, III; Vishnu Dutt, 'Problems of Pollution', *ToI*, 6 March 1972, 8; Sivadas Banerjee, 'Growing Pollution in West Bengal', *ToI*, 17 December 1975, 8.

[145] Centre for Science and Environment, *The State of India's Environment, 1982: A Citizens' Report* (New Delhi: Centre for Science and Environment, 1982).

[146] Centre for Science and Environment, *The State of India's Environment, 1984–85: The Second Citizens' Report* (New Delhi: Centre for Science and Environment, 1985).

[147] Emma Mawdsley, 'India's Middle Classes and the Environment', *Development and Change* 35 (2004): 79–103.

[148] 'Coal Power Plants Kill 120,000 People a Year', *Guardian*, 11 March 2013, 17.

Conclusion

Alien and intrusive, poison has a unique capacity to disturb and unsettle – even when it does not actually bare its fangs and kill. At its most insidious, poison blurs and disrupts the distinction between loyalty and treachery, between good and bad in food and medicine, between what nurtures and sustains and what is toxic to our minds and bodies. In society at large, poison, or the anxiety and dread to which it gives rise, can be equally unsettling. It can create a sense of vulnerability and distrust, threaten safety and security and even, on occasion, as a 'weapon of the weak', trouble the established regime – though we should not ignore the contrary possibility that poison might sometimes also be the embodiment of social violence and the means by which dissent is erased and the socio-political order upheld. The invisibility of poisons, and, in many instances, the uncertain identity of the poison and poisoner, only adds to the bewilderment and alarm poison can create. In India, as elsewhere, poison scares and panics have been a not infrequent phenomenon: they have stoked alarm and set off a search for scapegoats, but they have also been an incentive to science and an engine of governmental change. But it is undoubtedly the negatives about poison that prevail. As much by reason of its social use and cultural associations as from its chemical make-up and material effects, poison lends itself to harsh metaphor and critical moralizing. It evokes evil, brutality, betrayal. The world has known few works in praise of poison.

By looking at India since the early nineteenth century, this book has offered evidence of poison's universality. India, too, had its 'arsenic century'. India, too, had its sensational poison trials and infamous poisoners. But this discussion has also sought, in so far as a comparative judgement is possible, to suggest that India, historically, culturally and materially, has had a particularly long and singularly complex relationship with poison. Perhaps even among Asian societies, only China, with its ancient use of aconites and orpiment, its Western missionary condemnation of toxic remedies and its recent wholesale descent into industrial pollution, has a comparable tale to tell. But for India the poison presence was evident not only in myth and metaphor, but also in gender inequality, patriarchal dominance and caste supremacy. A notion of how poison might be countered or, alternatively, put to therapeutic use was embedded in Ayurveda, just as poison informed popular usage – in infanticide and abortion, in aphrodisiacs and elixirs – and gave a weighty duality to the very use and meaning of the term 'medicine'. To this mix of the cultural and the material, colonialism added still further layers of complexity. Colonial botanists, pharmacists and physicians argued for a starker dichotomy between poison and medicine than they believed India to possess but they also selectively added Indian poison plants and minerals to their own pharmacopoeia. Across society, from princes to highway robbers and cattle killers, poisoning served colonialism well – as a summation of the evils, the depravities, of India, just as it encapsulated a sense of European unease, of never quite being 'at home' in India.

But poison was also opportunity. In botany, medicine, chemistry, pharmacology and bacteriology, it helped impel and configure new fields of scientific endeavour – fields in which, over the long colonial century from the 1830s to the 1940s, both Europeans and Indians were active and creative participants. Investigation into poison's criminal and social uses stimulated the formation of new institutions and agencies – the chemical examiners and their forensic laboratories foremost among them. If poison troubled the colonial 'information order', the refinements of science were critical to quelling panic and restoring epistemic authority. As a representational trope but also, critically, as the harbinger of new forms of governance, poisoning influenced the emergence of a colonial view – sometimes crass, often subtle – of ethnology, criminology and medical jurisprudence, and, especially in the late nineteenth and early twentieth centuries, it helped foster the rise of an Indian public, as well as an imperial one. Toxicology, a science of the laboratory, was also a science of and for the public, even in the way in which the forensic evidence at its command might be so publicly questioned. Poison had no small part to play in the science, governance and social life of British India.

And yet the history of poisons – more exactly, the cumulative effect of the many stories told and retold about poison and worked up into broader narratives of dominance, subordination and defiance – did not end there. Poison took on ever-widening forms and functions, and in not all of these new incarnations did poison necessarily appear as an embodiment of evil. The controlled use of poison could be deployed (and on a huge scale) to eliminate unwanted dogs or destroy plague-bearing rats. Substances long regarded as poisons, their toxic potency redirected to therapeutic ends, could be used to attack some of the diseases that most troubled India and defined its tropical pathogenicity. Moreover, poison and poisoning were not the only ways in which toxicity came to manifest itself in British India, especially after the 1880s. To try to capture that larger social and environmental sense of toxicity, and to encapsulate for India what Lawrence Buell has so helpfully called 'fear of a poisoned world', it has been necessary to look beyond the immediate domain of poison itself, beyond the rise and fall of arsenic or *Aconitum ferox* as objects of distrust. Part of the potency of poison lies in its ability to reinvent itself. As one form of toxicity dies away, another – perhaps still worse – arises to take its place.

One way of addressing this – poison's wider penumbra – in relation to British India, and beyond to the post-colony, has been by considering how alarms over poisoned or polluted foods fuelled public panic and state concern in the nineteenth century but then morphed into growing apprehension about the contamination and adulteration of items of everyday purchase and consumption from milk to ghee to cooking oil. In the marketplace and in the home, poisoning was no less insidious or dangerous than the poisoning previously associated with homicidal murderers or low-caste cattle killers. In the leading cities of British India, too, poisoning began to manifest itself anew in the pollution of watercourses and reservoirs and in the polluted air that spewed from factory chimneys or lingered in the dust-laden atmosphere of textile mills, print works, paint shops and chemical factories. Here were new forms of poisoning – urban, commercial, industrial and environmental – in which culpability was still harder to identify and punish than in toxicity's earlier incarnations and in which the operations of the ill-regulated marketplace and the profit seeking of trade and industry were often deeply entrenched.

Pollution and poison were overlapping but never quite identical categories. Evidence of pollution, like that of poison, might evade the naked eye (and so likewise require the forensic skills of the bacteriologist and chemist), but it could also be highly visible. It might be unpleasant rather than actually toxic – a 'nuisance' in nineteenth-century parlance. Still more than poison, it might give rise to a complacent shrug and a degree of self-interested unconcern. Indeed, in India before 1947, many of the toxic side effects of urban industry – like lead poisoning – were ignored or, at best, insufficiently attended to. But pollution also has the capacity to poison on a monumental scale. Even if before Bhopal in 1984 there were few pollution episodes of such magnitude as to trigger poison scares, and few sensational trials of poisoners to compare with those of nineteenth- and early-twentieth-century poisoners, to galvanize the public and bestir the authorities into remedial action, there is no doubt that the widening tide of pollution demanded a redress it all too rarely received. In this 'new age of toxicity', pollution has become the new poison.

Bibliography

Archives

India Office Records, British Library, London

Bengal: Judicial, Medical, Municipal (Medical), Municipal (Sanitary), Public
Board of Control's Collections (F/4); L/P&J
Bombay: Public
India: Home (Judicial); Home (Sanitary); Legislative; Political; Public
Madras: Board of Revenue

National Archives of India, New Delhi

India: Home (Sanitary)

Tamil Nadu Archives, Chennai

Madras: Judicial; Local Self-Government (Public Health)

Wellcome Library, London

J. W. D. Megaw, Collected Papers

Official publications

Bombay Acts, 1862–70
Commission of Enquiry into Charges Laid against H. M. Mulharrao, Gaekwar of Baroda
Correspondence Relating to the Prohibition of Burials in the Back Bay Sands (Bombay: Bombay Education Society's Press, 1855)
Correspondence Relating to a Proposed Enactment for the Regulation of Places Used for the Disposal of Corpses in the Town and Island of Bombay (Bombay: Bombay Education Society's Press, 1855)
Indian Hemp Drugs Commission, vol. 5: Evidence of Witnesses from North-Western Provinces and Oudh and Punjab (Calcutta: Office of the Superintendent of Government Printing, India, 1894)

Poisons Manual, 1934 (Calcutta: Superintendent of Government Printing, Bengal, 1934)

Report of the Commission Appointed to Enquire into Certain Matters Connected with the Sanitation of the Town of Calcutta (Calcutta: Bengal Secretariat Press, 1885)

Report of the Commissioners Appointed to Inquire into the Origin, Nature Etc., of Indian Cattle Plagues (Calcutta: Office of the Superintendent of Government Printing, 1871)

Report of the Committee on the Indigenous Systems of Medicine (2 vols, Madras: Government Press, 1923)

Report of the Drugs Enquiry Committee, 1930–31 (Calcutta: Government of India, Central Publications Branch, 1931)

Report of Operations in the Thuggee and Dacoity Department during 1859 and 1860: Selections from the Records of the Government of India, Foreign Department, vol. 34 (Calcutta: Bengal Printing Company, 1861)

Report of the Smoke Abatement Committee, 1882 (London: Smith, Elder, 1883)

Report on the Effects of Artificial Respiration, Intravenous Injection of Ammonia and Administration of Various Drugs &c. in Indian and Australian Snake-Poisoning (Calcutta: Bengal Secretariat Press, 1874)

Royal Commission on Arsenical Poisoning: First Report of the Royal Commission Appointed to Inquire into Arsenical Poisoning from the Consumption of Beer and Other Articles of Food or Drink: Part I: Report (London: His Majesty's Stationery Office, 1901)

Selections from the Records of Government in the Police Branch of the Judicial Department, vol. 1 (Bombay: Bombay Education Society's Press, 1853)

Selections from Records of the Government of India, Home, Revenue, and Agricultural Department, vol. 167: *Papers Relating to the Crime of Robbery by Poisoning* (Calcutta: Office of the Superintendent of Government Printing, 1880)

Selections from the Records of Government, North-Western Provinces, vols 2, 4, 5 (Agra: Secundra Orphan Press, 1856)

Contemporary sources

Aberigh-Mackay, G. R., *The Chiefs of Central India* (Calcutta: Thacker, Spink, 1879)

Acton, Hugh W., 'The Causation of Epidemic Dropsy', *IMG* 57 (1922): 331–33
    'An Investigation into the Causation of Lathyrism in Man', *IMG* 57 (1922): 241–47
    and R. N. Chopra, 'The Nature and Pharmacological Action of Cholera Toxin', *IJMR* 12 (1924–25): 235–50.
    and R. N. Chopra, 'The Problem of Epidemic Dropsy and Beriberi', *IMG* 60 (1925): 1–18

Adam, H. L., *The Indian Criminal* (London: John Milne, 1909)

Adarkar B. P., *Report on Labour Conditions in the Chemical Industry* (Simla: Manager of Publications, Government of India, 1946)

Anderson, L. A. P., A. Howard and J. L. Simonsen, 'Studies on Lathyrism', *IJMR* 12 (1924–25): 613–44

Annesley, James, *Researches into the Causes, Nature, and Treatment of the Most Prevalent Diseases of India* (2 vols, London: Longman, Rees, Orme, Brown & Green, 1828)

[Anon.], *Autobiography of an Indian Army Surgeon* (London: Richard Bentley, 1854)

'The Chemical Examiner's Report (Bengal)', *IMG* 36 (1901): 303–04

'Current Topics', *IMG* 36 (1901): 71

'The First Indian Medical Congress', *British Medical Journal*, 9 February 1895, 310–13

'Notes from India', *Lancet* 178, 2 September 1911, 731–32

'Poisoning in India', *British Medical Journal*, 17 September 1892, 641–42

'The Poisons Used to Destroy Human Life in Bengal', *IMG* 20 (1885): 320–21

The Trial of Shama Charan Pal: An Illustration of Village Life in Bengal (London: Lawrence & Bullen, 1897)

Arthur, T. C., *Reminiscences of an Indian Police Official* (London: Sampson, Low, Marston, 1894)

[Babur], *Babur Nama: Journal of Emperor Babur* (New Delhi: Penguin, 2006)

Bagchi, K. N., 'Incidence of Lead Poisoning among Hindu Women and Children', *IMG* 76 (1941): 23–29

and H. D. Ganguly, 'Arsenic in Food', *IMG* 76 (1941): 720–22

and H. D. Ganguly, 'Arsenic in Human Tissues and Excreta', *IMG* 72 (1937): 477–81

and H. D. Ganguly, 'Lead in Urine and Faeces', *IJMR* 25 (1937–38): 147–54

H. D. Ganguly and J. N. Sirdar, 'Lead Content of Human Hair', *IJMR* 27 (1939–40): 777–91

H. D. Ganguly and J. N. Sirdar, 'Lead in Human Tissues', *IJMR* 26 (1938–39): 935–46

Balfour, Edward, *The Cyclopaedia of India* (3rd ed., 3 vols, London: Bernard Quaritch, 1885)

Ball, V., *A Manual of the Geology of India, Part III: Economic Geology* (Calcutta: Office of the Geological Survey of India, 1881)

Banerjea, R. and A. K. Sen, 'The Bacterial Content of the Calcutta Milk Supply', *IMG* 81 (1946): 40–45a

Banerjee, Nil Rattan, 'The Symptoms in Datura-Poisoning', *IMG* 20 (1885): 209–11

Bannerman, W. B., 'Note on Arsenic as a Prophylactic for Malaria', *IMG* 26 (1891): 70–71

Barnes, F. D., 'Problems Relating to Working Mothers and Infants', *Social Service Quarterly* 8 (1922): 12–15

Barry, Collis, *Legal Medicine (in India) and Toxicology* (2nd ed., 2 vols, Bombay: Thacker, 1904)

Basu, B. D., 'On the Study of Indigenous Drugs', *IMG* 28 (1893): 336–38

Baynes, C. R., *Hints on Medical Jurisprudence, Adapted and Intended for the Use of Those Engaged in Judicial and Magisterial Duties in British India* (Madras: Pharoah, 1854)

Beck, Theodric Romeyn and John R. Beck, *Elements of Medical Jurisprudence* (5th ed., London: Longman, Rees, Orme, Brown, Green & Longman, 1836)

Bedford, C. H., 'Notes on Some Toxicological Experiences in Bengal and in the Punjab', *IMG* 37 (1902): 202–07

Bellasis, A. F. (comp.), *Reports of Criminal Cases Determined in the Court of Sudder Foujdarree Adawlut of Bombay* (Bombay: Government Press, 1849)

Bentley, Arthur J. M., *Beri-Beri: Its Etiology, Symptoms, Treatment, and Pathology* (Edinburgh: Young J. Pentland, 1893)

Bernier, François, *Travels in the Mughal Empire, A. D. 1656–1668* (trans. Irving Brock, Westminster: Archibald Constable, 1891)

Bhaskaran, T. R., 'A Plea for Water Pollution Research', *IMG* 82 (1947): 750–52

Bhuttacharjee, Hem Chunder, 'Case of Poisoning by Gloriosa Superba', *IMG* 7 (1872): 153

Birch, Edward A., *The Management and Medical Treatment of Children in India* (4th ed., Calcutta: Thacker, Spink, 1902)

Blaney, Thomas, 'Brief Notes on Enteric Fever', *IAMS* 10 (1870): 351–62

Blyth, Alexander Wynter and Meredith Wynter Blyth, *Poisons: Their Effects and Detection* (4th ed., London: Charles Griffin, 1906)

Bose, A. C., 'A Modified Method of Estimating Arsenic-Content of Indian Food-Stuffs', *IJMR* 22 (1935): 697–700

Bose, A. N., T. N. Ghosh, S. N. Mitra and S. Dutta, 'On the Toxicity of Some Organic Antimonial Drugs Used for the Treatment of Kala-Azar', *IMG* 81 (1946): 13–16

Bose, Chunilal, 'The Bhowanipore Food-Poisoning Case', in Bose, *Scientific Papers* 1: 298–333

   'A Brief Survey of Research-Work in Chemistry in Bengal', in Bose, *Scientific Papers* 1: 104–12

   'A Few Hints on Sanitary Reconstruction', in Bose, *Scientific Papers* 2: 175–91

   Food (Calcutta: University of Calcutta, 1930)

   'The Milk-Supply of Calcutta: Its Hygienic, Commercial and Social Aspects', in Bose, *Scientific Papers* 2: 113–60

   'On the Chemistry and Toxicology of *Nerium Odorum*', *IMG* 36 (1901): 287–90

   'The Toxic Principles of the Fruits of *Luffa aegyptiaca*', in Bose, *Scientific Papers* 1: 86–103

Bose, J. P. (ed.), *The Scientific and Other Papers of Rai Chunilal Bose Bahadur* (2 vols, Calcutta: Forward Press, 1924)

Bourchier, George, *Eight Months' Campaign against the Bengal Sepoy Army during the Mutiny of 1857* (London: Smith, Elder, 1858)

Boyd, T. C., L. Everard Napier and A. C. Roy, 'The Distribution of Antimony in the Body Organs', *IJMR* 19 (1931): 285–94

Brahmachari, Upendranath, *Gleanings from My Researches* (2 vols, Calcutta: University of Calcutta, 1940–41)

Brown, J. Coggin, *Notes on Antimony, Arsenic and Bismuth* (Calcutta: Superintendent, Government Printing, India, 1921)

Brown, T. E. B., *Punjab Poisons* (3rd ed., Lahore: 'Civil and Military Gazette' Press, 1888)

Browne, G. Lathom and C. G. Stewart, *Reports of Trials for Murder by Poisoning by Prussic Acid, Strychnia, Antimony, Arsenic, and Aconita* (London: Stevens & Sons, 1883)

Brunton, T. Lauder and J. Fayrer, 'On the Nature and Physiological Action of the Poison of *Naja tripudians* and Other Indian Venomous Snakes', *Proceedings of the Royal Society of London* 21 (1872–73): 358–74

Buchanan, Andrew, *Report on Lathyrism in the Central Provinces in 1896–1902* (Nagpur: Albert Press, 1908)

Buchanan, Francis, *An Account of the Districts of Bihar and Patna in 1811–12* (2 vols, Patna: Bihar and Orissa Research Society, n.d.)

[Buchanan] Francis Hamilton, *An Account of the Kingdom of Nepal* (Edinburgh: Archibald Constable, 1819)

Buchanan, W. J., 'A Chapter on Medical Jurisprudence in India', in Fred. J. Smith (ed.), *Taylor's Principles and Practice of Medical Jurisprudence* (7th ed., 2 vols, London: J. & A. Churchill, 1920), 2: 886–920

Burnett-Hurst, A. R., *Labour and Housing in Bombay: A Study in the Economic Conditions of the Wage-Earning Classes in Bombay* (London: P. S. King, 1925)

Buxton, P. A., 'The Use of the New Insecticide DDT in Relation to the Problem of Tropical Medicine', *Transactions of the Royal Society of Tropical Medicine and Hygiene* 38 (1945): 367–93

Caius, J. F. and K. S. Mhaskar, 'The Correlation between the Chemical Composition of Anthelmintics and Their Therapeutic Values in Connection with the Hookworm Inquiry in the Madras Presidency', *IJMR* 7 (1919–20): 429–63

Campbell, Mr Justice, 'The Ethnology of India', *JASB* 35 (1866): 1–152

Campos, J. J., 'Chronic Lead Poisoning in the Printing Presses of Calcutta', *IMG* 56 (1921): 175–78

Candy, R. H., 'A Note on the Prevalence of Lead Poisoning in India', *IMG* 68 (1933): 136–37

Caraka Samhita (6 vols, Jamnagar: Shree Gulabkunverba Ayurvedic Society, 1949)

Caraka Samhita (5 vols, Delhi: Sri Satguru Publications, 1996)

Chakraborty, M. K., M. N. Rao and B. Banerji, 'A Study of the Occupational Lead Hazard in Selected Indian Industries', *IJMR* 38 (1950): 429–56

Chatterjee, Bankim Chandra, *The Poison Tree: A Tale of Hindu Life in Bengal* (London: T. Fisher Unwin, 1884)

Chatterton, Alfred, *Monograph on Tanning and Working in Leather in the Madras Presidency* (Madras: Superintendent, Government Press, 1905)

Chevers, Norman, *A Commentary on the Diseases of India* (London: J. & A. Churchill, 1886)

A Manual of Medical Jurisprudence for Bengal and the North-Western Provinces of India (2nd ed., Calcutta: Bengal Military Orphan Press, 1856)

A Manual of Medical Jurisprudence for India (3rd ed., Calcutta: Thacker, Spink, 1870)

'Report on Medical Jurisprudence in the Bengal Presidency', *IAMR* 2 (1854): 243–426

Chopra, R. N., *Indigenous Drugs of India: Their Medical and Economic Aspects* (Calcutta: N. Mukherjee, 1933)

R. L. Badhwar and S. Ghosh, *Poisonous Plants of India* (2 vols, New Delhi: Indian Council of Agricultural Research, 1965)

and N. N. Ghose, 'Addiction to "Post" – Unlanced Capsules of *Papaver somniferum*: Part II', *IJMR* 19 (1931): 415–21

Christison, Robert, *A Treatise on Poisons* (4th ed., Philadelphia: Edward Barrington and George D. Haswell, 1845)

Church, A. H., *Food-Grains of India* (London: Chapman & Hall, 1886)

'Vichka Seed as a Famine Food in the Bombay Presidency', *Agricultural Ledger* 6 (1899): 1–2

Cleghorn, J., 'Cases of Datura Poisoning', *IMG* 6 (1871): 209

Clemensha, William Wesley, *Sewage Disposal in the Tropics* (Calcutta: Thacker, Spink, 1910)

Cook, J. Neild, 'The Bhowanipore Food-Poisoning', *IMG* 38 (1903): 362–64

Cornish, William Robert, *Reports on the Nature of the Food of the Inhabitants of the Madras Presidency* (Madras: United Scottish Press, 1863)

Cox, Edmund, *Police and Crime in India* (London: Stanley Paul, 1911)

Crawford, D. G., *A History of the Indian Medical Service, 1600–1913* (2 vols, London: W. Thacker, 1914)

Cullen, J. P., 'A Fatal Case of Poisoning by Neo-Karsivan', *IMG* 59 (1924): 245

Cunningham, D. D., 'The Physiological Action of Cobra-Venom', in *Scientific Memoirs by Medical Officers of the Army of India, Part XI* (Calcutta: Office of the Superintendent of Government Printing, India, 1898)

Cunningham, J. A., 'A Note on the Suppression of Cholera in a Famine Camp', *IMG* 35 (1900): 385–86

Curry, J. C., *The Indian Police* (London: Faber & Faber, 1932)

Da Orta, Garcia, *Colloquies on the Simples and Drugs of India* (London: Henry Sotheran, 1913)

Das Gupta, S. C. (ed.), *The Bhowal Case: High Court Judgements* (Calcutta: S. C. Sarkar & Sons, 1941)

Dennys, George W. P., 'Iron and Arsenic as a Cure for and a Prophylactic against Malaria', *IMG* 51 (1916): 242–46

De Quincey, Thomas, *Confessions of an English Opium Eater* (London: Penguin, 1971)

On Murder (Oxford: Oxford University Press, 2006)

Dey, Kanny Lall, *The Indigenous Drugs of India* (2nd ed., Calcutta: Thacker, Spink, 1896)

'Medicinal Substances Used by Native Practitioners', in A. M. Dowleans (ed.), *Official Classified and Descriptive Catalogue of the Contributions from India to the London Exhibition of 1862* (Calcutta: Bengal Printing Co., 1862), 65–81

Doyle, Arthur Conan, *The Adventures of Sherlock Holmes* (London: Penguin, 1994)

Dumas, Alexandre, *The Count of Monte Cristo* (London: Penguin Books, 2003)

Dutt, Udoy Chand, *The Materia Medica of the Hindus* (Calcutta: Thacker, Spink, 1877)

Dymock, William, *Pharmacographia Indica: A History of the Principal Drugs of Vegetable Origin, Met With in British India* (3 vols, Calcutta: Thacker, Spink, 1890–91)

Editorial, 'Cholera in the Port', *IMG* 1 (1866): 190–91
    'The Crusade against Opium', *IMG* 27 (1892): 178–80
    'Dr H. E. Durham's Report on Beriberi', *IMG* 39 (1904): 221–22
    'Industrial Medicine and Hygiene in Bengal', *IMG* 56 (1921): 182
    'Lathyrism', *IMG* 74 (1939): 421–22
    'The Materia Medica of the Hindus', *IMG* 12 (1877): 189–90
    'Medical Jurisprudence in India', *IMG* 24 (1889): 309–10
    'The Opium Commission', *IMG* 29 (1894): 21
    'Our Special Medico-Legal Number', *IMG* 37 (1902): 201–02
    'A Plea for Hakeems', *IMG* 3 (1868): 87–89
    'Poisoning by Antimony', *IMG* 38 (1903): 383–84
    'The Recent Calcutta Ghi-Adulteration Case', *IMG* 36 (1901): 301–03
    'River Pollution in India', *IMG* 25 (1890): 56–57
    'Strychnia Poisoning in India', *IMG* 20 (1885): 76–77
Edwardes, S. M. *Crime in India* (London: Oxford University Press, 1924)
Elliot, H. M. and John Dowson, *The History of India, as Told by Its Own Historians: The Muhammadan Period, Vol. II* (London: Trübner, 1869)
Evans, J. F. and Chunilal Bose, 'The Necessity for an Act Restricting the Free Sale of Poisons in Bengal', in *Transactions of the First Indian Medical Congress*, 467–87
Farquhar, J. N., *Modern Religious Movements in India* (New York: Macmillan, 1915)
Fayrer, J., 'A Case of Acute Malarial Poisoning', *IMG* 26 (1891): 296–301
    'Deaths from Snake-Bites', *IMG* 5 (1870): 156–57
    'Destruction of Life in India by Poisonous Snakes', *Nature*, 28 December 1882, 205–08
    'Destruction of Life in India by Wild Animals', *Nature*, 13 January 1883, 268–70
    'Note on the Use of Snake-Poison in Medicine, by the Koberajes of Bengal', *IAMS* 14 (1870): 226–31
Fitzpatrick, P., 'Case of Oleander Poisoning', *IMG* 24 (1889): 307
Fleming, John, 'A Catalogue of Indian Medicinal Plants and Drugs', *AR* 11 (1810): 153–96
Forbes, James, *Oriental Memoirs: A Narrative of Seventeen Years Residence in India* (2nd ed., 2 vols, London: Richard Bentley, 1834)
Fryer, John, *A New Account of East India and Persia in Eight Letters* (London: Richard Chiswell, 1698)
[Gandhi, M. K.], *Collected Works of Mahatma Gandhi* 13 (New Delhi: Publications Division, Ministry of Information and Broadcasting, 1964)
    Gandhi's Speeches and Writings (Madras: G. A. Natesan, n.d.)
Ghose, Sarat Chandra, *Beri-Beri: Its Causation, Prevention and Homoeopathic Treatment* (Calcutta: A. C. Dutt, 1910)
Giles, A. H., 'Poisoners and Their Craft', *CR* 81 (1885): 78–122
Gimlette, John D., *Malay Poisons and Charm Cures* (3rd ed., Kuala Lumpur: Oxford University Press, 1971)

Gladwin, Francis, *Ulfaz Udwiyeh, or the Materia Medica in the Arabic, Persian, and Hidevy Languages Compiled by Noureddeen Mohammed Abdullah Shirazy* (Calcutta: Chronicle Press, 1793)

Gopalakrishnan, V. R., 'Cattle Poisoning in Assam', *Indian Farming* 5 (1944): 77–79

Gore, S. N., 'Calcium Cyanide Fumigation', *IJMR* 13 (1925–26): 287–99

Graham, J. D., 'Medical and Research Organisation', in Far Eastern Association of Tropical Medicine, *The Indian Empire* (Calcutta: Thacker, 1927), 81–108

Greave, Peter *The Seventh Gate* (Harmondsworth: Penguin, 1978)

Greene, J. A., 'Case of Poisoning by Dhatoora', *IMG* 6 (1871): 165

Greig, E. D. W., *Epidemic Dropsy in Calcutta (Final Report)* (Calcutta: Superintendent of Government Printing, India, 1912)

Gribble, J. D. B. and Patrick Hehir, *Outlines of Medical Jurisprudence for India* (4th ed., Madras: Higginbotham, 1898)

Grover, Frederick, *Report on the Abatement of Smoke Nuisance in Calcutta* (Simla: Government Central Printing Office, 1903)

Gubbins, Martin Richard, *An Account of the Mutinies in Oudh, and of the Siege of the Lucknow Residency* (London: Richard Bentley, 1858)

Gunthorpe, E. J., *Notes on Criminal Tribes Residing in or Frequenting the Bombay Presidency, Berar and the Central Provinces* (Bombay: 'Times of India' Steam Press, 1882)

Gupta, Sen and Nagendra Nath, *The Ayurvedic System of Medicine* (2 vols, Calcutta: Keval Ram Chatterjee, 1901)

Guy, William A. and David Ferrier, *Principles of Forensic Medicine* (7th ed., London: Henry Renshaw, 1895)

Haffkine Institute Platinum Jubilee Commemoration Volume, 1899–1974 (Bombay: Haffkine Institute, n.d.)

Hankin, E. H., 'Directions for the Use of Permanganate of Potassium in Combating Water-Borne Diseases', *IMG* 31 (1896): 241–47

The Mental Limitations of the Expert (2nd ed., Calcutta: Butterworth, 1921)

Harvey, Robert, 'Report on the Medico-Legal Returns Received from the Civil Surgeons in the Bengal Presidency during the Years 1870, 1871, and 1872', *IMG* 11 (1876): 57–62, 85–89, 113–19, 141–46, 169–74, 197–201

Hehir, Patrick, *Hygiene and Diseases of India: A Popular Handbook* (3rd ed., Madras: Higginbothams, 1913)

Opium: Its Physical, Moral, and Social Effects (London: Ballière, Tindall & Cox, 1894)

Hervey, H , *Cameos of Indian Crime* (London: Stanley Paul, 1912)

Hollins, S. T., *No Ten Commandments: Life in the Indian Police* (London: Hutchinson, 1954)

Holmes, J. D. E., *A Note on Some Interesting Results Following the Internal Administration of Arsenic in Canker and Other Diseases of the Foot in Horses* (Calcutta: Superintendent, Government Printing, India, 1912)

Honigberger, John Martin, *Thirty-Five Years in the East* (London: H. Ballière, 1852)

Hume, T., 'Case of Partial Paralysis, Supposed to Have Followed the Injudicious Administration of Arsenic', *IMG* 11 (1876): 103

Irvine, R. H., *A Short Account of the Materia Medica of Patna* (Calcutta: Military Orphan Press, 1848)

Irving, James, 'Report on a Species of Palsy Prevalent in Pergunnah Khyragurh, in Zillah Allahabad, from the Use of Kessaree Dall, as an Article of Food', in *Selections from the Records of Government, North-Western Provinces, Vol. 2* (Allahabad: Government Press, North-Western Provinces, 1866), 265–76

'Notice of Paraplegia Caused by the Use of Lathyrus Sativus in the Various Districts of the North-Western Provinces of India', *IAMS* 12 (1868): 89–124

Iyar, Ranganadha, *Dr. S. Swaminadhan: A Memoir* (Madras: Hoe, n.d.)

Johnston, J. W., 'Poisoning by Repeated Small Doses of Arsenic', *IMG* 8 (1873): 185–86

Jones, J. A., *A Manual of Hygiene, Sanitation and Sanitary Engineering with Special Reference to Indian Conditions* (Madras: Superintendent, Government Press, 1896)

Kalra, S. L., V. P. Jacob and K. N. A Rao, 'Effect of DDT and BHC on *Ornithodoros* Ticks', *IJMR* 38 (1950): 457–66

[Kaye, John and G. B. Malleson], *Kaye's and Malleson's History of the Indian Mutiny of 1857–8* (6 vols, London: W. H. Allen, 1889–93)

Kitts, Eustace J., *Serious Crime in an Indian Province* (Bombay: Education Society's Press, 1889)

Kunhardt, J. G. C., 'The Rat Problem in India', *IJMR*, special issue (1919): 145–72

Lal, R. B. and A. C. Das Gupta, 'Investigations into Epidemic Dropsy, Part X', *IJMR* 29 (1941): 157–65

and S. C. Roy, 'Investigations into the Epidemiology of Epidemic Dropsy, Part I', *IJMR* 25 (1937–38): 163–76

Lambert, D. P., *The Medico-Legal Post-Mortem in India* (London: J. & A. Churchill, 1937)

Latham, R. G., *Ethnology of India* (London: John van Voorst, 1859)

Lely, F. S. P., *Suggestions for the Better Governing of India* (London: Alston Rivers, 1906)

Lisboa, J. C., 'Famine Plants: Wild Herbs, Tubers, etc. Used as Food during Seasons of Scarcity', in *Gazetteer of the Bombay Presidency, Vol. 25: Botany* (Bombay: Government Central Press, 1886), 190–209

Liston, W. Glen, '"The Next War": Man *versus* Insects', *IJMR*, special issue (1919): 18–25

'Plague, Rats, and Fleas', *IMG* 40 (1905): 43–49

'The Use of Hydrocyanic Gas for Fumigation', *IJMR* 7 (1919–20): 778–802

and S. N. Gore, "Abstract of a Paper on Hydrocyanic Acid Gas as an Insecticide', *IJMR* special issue (1919): 40–42

Lobo-Mendonca, R., 'A Note on Paranitraniline Poisoning', *IMG* 77 (1942): 673–75

Lyon, I. B., *A Text Book of Medical Jurisprudence for India* (Calcutta: Thacker, Spink, 1889)

Maclean, William Campbell, *Diseases of Tropical Climates* (London: Macmillan, 1886)

Maconochie, Evan, *Life in the Indian Civil Service* (London: Chapman & Hall, 1926)

Mair, R. S., *Statistics of Unnatural Deaths in Madras and Other Presidencies and Provinces of India* (Madras: Gantz Brothers, 1868)

Mall, G. D., ' "Sui" or Needle Poisoning in the Punjab in Cattle', in *Transactions of the First Indian Medical Congress*, 503–05

Mann, Thomas, *The Magic Mountain* (London: Vintage Books, 1999)

Marshman, John Clark, *The History of India* (3 vols, London: Longmans, Green, Reader & Dyer, 1867)

Martin, James Ranald, *The Influence of Tropical Climates on European Constitutions* (London: John Churchill, 1856)

Maxwell-Lefroy, H., *Indian Insect Pests* (Calcutta: Office of the Superintendent of Government Printing, India, 1908)

Mayne, T., 'Opium-Eaters', *IMG* 16 (1881): 89–90

McLeod, Kenneth, *Medico-Legal Experience in the Bengal Presidency, Being a Report on the Medico-Legal Returns Received from the Civil Surgeons of Bengal during the Years 1868 and 1869* (Calcutta: Central Press, 1875)

McReddie, G. D., 'Opium Suicides in Hardoi District', *IMG* 26 (1891): 168–69

Megaw, J. W. D., 'The Beriberi Problem', in J. W. D. Megaw, *Collected Papers* (Wellcome Library, London)

'Notes on Cases of the "Epidemic Dropsy" Type of Beri-Beri at the Presidency General Hospital, Calcutta', in J. W. D. Megaw, *Collected Papers* (Wellcome Library, London), 45 (1910)

S. P. Bhattacharji and B. K. Paul, 'Further Observations on the Epidemic Dropsy Form of Beri-Beri', *IMG* 63 (1928): 417–39

Mehrotra, K. N., 'Use of DDT and Its Environmental Effects in India', *Proceedings of the Indian National Science Academy* B51 (1985): 169–84

Menpes, Mortimer and Flora Annie Steel, *India* (London: Adam & Charles Black, 1905)

Modi, Jaising P., *A Text-Book of Medical Jurisprudence and Toxicology* (2nd ed., Calcutta: Butterworth, 1922)

Modi, Jivanji Jamshedji, 'The Vish Kanya or Poison Damsel of Ancient India', in *Anthropological Papers, Vol. 4* (Bombay: British India Press, 1929), 226–39

Moor, Edward, *Hindu Infanticide: An Account of the Measures Adopted for Suppressing the Practice of the Systematic Murder by Their Parents of Female Infants* (London: J. Johnson, 1811)

Moore, W. J., *A Manual of Family Medicine for India* (4th ed., London: J. & A. Churchill, 1883)

'The Opium Question', *IMG* 16 (1881): 211–15, 265–69

The Other Side of the Opium Question (London: J. & A. Churchill, 1882)

Mouat, Frederic J., *Reports on Jails Visited and Inspected in Bengal, Behar, and Arracan* (Calcutta: Military Orphan Press, 1856)

Mukhopadhyaya, Girindranath, *History of Indian Medicine from the Earliest Ages to the Present Time* (2 vols, Calcutta: University of Calcutta, 1923)

Mukhtar, Ahmad, *Report on Labour Conditions in Tanneries and Leather Goods Factories* (Simla: Manager, Government of India Press, 1946)

Murray, T., 'Case of Poisoning from the Oleander Root', *IMG* 12 (1877): 319–20

Nicholson, William, *Report on Smoke Nuisances and Their Abatement in Calcutta* (Calcutta: Bengal Secretariat Book Depot, 1906)

Oman, John Campbell, *Cults, Customs and Superstitions of India* (London: T. Fisher Unwin, 1908)

Orfila, M. P., *A General System of Toxicology, or Treatise on Poisons* (London: E. Cox & Son, 1816)

O'Shaughnessy, W. B., *The Bengal Dispensatory* (Calcutta: W. Thacker, 1842)
  Bengal Dispensatory and Pharmacopoeia (Calcutta: Bishop's Press, 1841)
  The Bengal Pharmacopoeia (Calcutta: Bishop's College Press, 1844)
  'On the Detection of Arsenic Poisons by Marsh's Process', *JASB* 8 (1839): 147–49
  Report on the Investigation of Cases of Real or Supposed Poisoning (Calcutta: Bishop's College Press, 1841)

Owen, Eliza, 'Introduction' to [Anon.], *The Trial of Shama Charan Pal: An Illustration of Village Life in Bengal* (London: Lawrence & Bullen, 1897), v–x

Owen, William, 'Report on the Treatment of Acute Dysentery by Aconite', *IMG* 17 (1882): 90–95

Pandalai, K. G., 'A Case of Salvarsan Poisoning', *IMG* 49 (1914): 59

Parks, Fanny, *Wanderings of a Pilgrim in Search of the Picturesque* (2 vols, London: Pelham Richardson, 1850)

Pearse, William H., 'The Prophylactic Use of Arsenic and Quinine against Cholera', *IMG* 17 (1882): 190–91

Pearson, Francis, *Memories of a K.C.'s Clerk* (London: Sampson, Low, Marston, n.d. [c. 1935])

Pillai, S. Chidambara Thanu, *Siddha System of Toxicology* (Madras: Siddha Medical Literature Research Centre, 1993)

Playfair, George, *The Taleef Shereef, or Indian Materia Medica* (Calcutta: Medical and Physical Society of Calcutta, 1833)

Postans, [Marianne] *Cutch; or, Random Sketches, Taken during a Residence in One of the Northern Provinces of Western India* (London: Smith, Elder, 1839)

Puri, I. M. and Rajindar Pal, 'Studies of Some Insecticides against Anopheles Adults and Larvae', *IJM* 1 (1947): 133–58

Raikes, Charles, *Notes on the North-Western Provinces of India* (London: Chapman & Hall, 1852)

Raju, V. Govinda, 'Deep Tube-Well Water of Bengal', *IJMR* 19 (1931–32): 447–55

Ramsay, H. M., *Detective Footprints: Bengal, 1874–1881* (London: Army and Navy Co-Operative Society, 1882)

Rankine, Robert, *Notes on the Medical Topography of the District of Sarun* (Calcutta: G. H. Huttmann, 1839)

Ray, P. C., *Life and Experiences of a Bengali Chemist* (Calcutta: Chuckerbutty, Chatterjee, 1932)

Ray, Rames Chandra, *Outlines of Medical Jurisprudence and Treatment of Poisoning* (2nd ed., Calcutta: The Hare Pharmacy, 1912)

Reinherz, O., 'The Seeds of *Shorea robusta* as a Famine Food', *Agricultural Ledger* 11 (1904): 33–36

Richards, V., 'Criminal Abortion', *IMG* 6 (1871): 230–31

Roberts, William, *Collected Contributions on Digestion and Diet* (2nd ed., London: Smith, Elder, 1897)

Roux, F., 'Arsenic in the Treatment of Kala-Azar', *IMG* 48 (1913): 132–33

Roy, G. C., 'Observations on the Nature of Cholera Poison', *IMG* 8 (1873): 120–22

Roy, Tarra Prosonno, 'On the Use of Dhatura as a Mydriatic', *IMG* 5 (1870): 187–88

Royle, J. Forbes, *Illustrations of the Botany and Other Branches of the Natural History of the Himalayan Mountains, and of the Flora of Cashmere* (2 vols, London: W. H. Allen, 1839)

A Manual of Materia Medica and Therapeutics (3rd ed., London: John Churchill, 1856)

Rudolf, Norman S., 'Notes on the Chemistry of the Seeds of Arbus Precatorius and "Sutari" Poisoning', in *Transactions of the First Indian Medical Congress, Calcutta* (Calcutta: Caledonian Steam Printing Works, 1895), 487–88

Russell, A.J.H. (ed.), *McNally's Sanitary Handbook for India* (6th ed., Madras: Superintendent, Government Press, 1923)

Russell, Patrick, *An Account of Indian Serpents Collected on the Coast of Coromandel* (4 vols, London: G. Nicol, 1795–1809)

Sabnis, C. V., 'Evaluation of Lead Hazard in a Pigment Manufacturing Concern', *IJMR* 40 (1952): 53–62

Sarkar, S. C., *Notable Indian Trials* (3rd ed., Calcutta: M. C. Sarkar & Sons, 1962)

Sarkar, Sarasi Lal, 'The Action of Quinine and Arsenical Preparations in Kala-Azar', *IMG* 50 (1915): 92–94

Seal, S. C. and M. N. De, 'Epidemic Dropsy', *Indian Medical Record* 58 (1937): 120–28

Sen, Binod Lall and Athutosh Sen, 'Preface', to Udoy Chand Dutt, *The Materia Medica of the Hindus* (2nd ed., 1900, reprinted Delhi: Mittal Publications, 1989), i–v

Seth, G. K. and T. R. Bhaskaran, 'Effect of Industrial Wastes Disposal on the Sanitary Condition of the Hooghly River in and Around Calcutta', *IJMR* 38 (1950): 341–56

Sherwood, Richard C., 'Of the Murderers Called Phansigars', *AR* 13 (1820): 250–81

Shourie, K. L., 'An Outbreak of Lathyrism in Central India', *IJMR* 33 (1945–46): 239–48

Simms, F. W., *Report on the Establishment of Water-Works to Supply the City of Calcutta* (Calcutta: Military Orphan Press, 1853)

Singh, Dulip, 'Modes of Inducing Criminal Abortion in the Punjab', *IMG* 20 (1885): 8–10

Sinhjee, Bhagavat, *A Short History of Aryan Medical Science* (London: Macmillan, 1896)

Sircar, Jadub Kristo, 'A Case of Morphia Poisoning', *IMG* 14 (1879): 259

Sircar, Mahendralal, *A Sketch of the Treatment of Cholera* (2nd ed., Calcutta: P. Sircar, 1904)

Skipton, G., 'Three Cases Shewing the Beneficial Effects of Dhatura in Asthma', *TMPSC* 1 (1825): 121–23

Sleeman, W. H., *Ramaseeana, or a Vocabulary of the Peculiar Language Used by the Thugs* (Calcutta: Military Orphan Press, 1836)

Sleeman, W. H., *Rambles and Recollections of an Indian Official* (2 vols, London: J. Hatchard & Sons, 1844)

Sleeman, W. H., *Report on Budhuk alias Bagree Dacoits and Other Gang Robbers by Hereditary Profession and on the Measures Adopted by the Government of India for Their Suppression* (Calcutta: Bengal Military Orphan Press, 1849)

Smith, D. B., 'Deaths from Coal Gas', *IMG* 7 (1872): 76–77

Smith, David R., 'Suspected Criminal Poisoning by Dhatoora in the Person of a European', *IMG* 3 (1868): 58–60

Smith, S. Browning, 'Rat Destruction and Plague', in *Supplement to the Indian Journal of Medical Research: Proceedings of the Third All-India Sanitary Conference* (5 vols, Calcutta: Thacker, Spink, 1914), 5: 158–61

Sokhey, S. S., G. D. Chitre and S. K. Gokhale, 'The Relative Value of Some Proprietary Cyanide Preparations for the Extermination of Rats and Fleas as a Plague-Preventive Measure', *IJMR* 27 (1939–40): 389–407

Somerville, Augustus, *Crime and Religious Beliefs in India* (Calcutta: The Criminologist, 1929)

Spry, Henry, *Modern India, with Illustrations of the Resources and Capabilities of Hindustan* (2 vols, London: Whittaker, 1837)

Srivastava, R. S., 'Malaria Control Measures in the Tarai Area under the Tarai Colonization Scheme', *IJM* 4 (1950): 151–65

Stapf, Otto, *Annals of the Royal Botanic Garden, Calcutta, Vol. X, Part II: The Aconites of India* (Calcutta: Bengal Secretariat Press, 1905)

Strong, F. P., *Extracts from the Topography and Vital Statistics of Calcutta* (Calcutta: no publisher, 1837)

Subrahmanyam, K., T. R. Bhaskaran and C. Chandra Sekar, 'Studies on Rural Water-Supplies', *IJMR* 36 (1948): 211–47
  and N. Majumdar, 'Environmental Conditions within Jute Mills', *IJMR* 39 (1951): 595–623

Subramanya Aiyar, N., 'Certain Facts Regarding the Poison-Lore of the Hindus', *IMG* 31 (1896): 5–8

Sundara Ayyar, K. V. and K. Narayanaswami, 'Varagu Poisoning', *Nature* 163 (1949): 912–13

Swarup, Anand, 'Acute "Kodon" Poisoning', *IMG* 57 (1922): 257–58

Tagore, Rabindranath, *Selected Short Stories* (London: Macmillan, 1991)

Taylor, Alfred S., *On Poisons, in Relation to Medical Jurisprudence and Medicine* (London: John Churchill, 1848)

Thomas, D. R., 'Cases of Poisoning and Suspected Poisoning', *IMG* 76 (1941); 429

[The Times], *India and the Durbar* (London: Macmillan, 1911)

Tod, James, *Annals and Antiquities of Rajasthan* (3 vols, London: Oxford University Press, 1920)

Transactions of the First Indian Medical Congress, Calcutta, 24th to 29th December, 1894 (Calcutta: Caledonian Steam Printing Works, 1895)

[Trevelyan, C. E.], 'The Thugs, or Secret Murderers of India', *Edinburgh Review* 64 (1837): 357–95

Trumpp, Ernest (ed.), *The Adi Granth, or the Holy Scriptures of the Sikhs* (London: William Allen, 1877)

Tulloch, Hector, *The Water-Supply of Bombay* (London: W. J. Johnson, 1872)

Turner, J. A., *Sanitation in Bombay* (Bombay: 'Times of India', 1914)

Varis, S. M., *A Study of Malaria and Beri-Beri* (Allahabad: Pioneer Press, 1912)

Venkata Swamy, J., 'Poisoning by Strychnos Nux Vomica', *IMG* 24 (1889): 113

Viswanathan, D. K., *The Conquest of Malaria in India: An Indo-American Co-Operative Effort* (Madras: Company Law Institute Press, 1958)

and T. Ramachandra Rao, 'Control of Rural Malaria with DDT Indoor Residual Spraying in Kanara and Dharwar Districts, Bombay Province', *IJM* 1 (1947): 503–42

Waddell, L. A., *Lyon's Medical Jurisprudence for India* (5th ed., Calcutta: Thacker, Spink, 1914)

Wall, A. J., *Indian Snake Poisons, Their Nature and Effects* (London: W. H. Allen, 1883)

Wall, F., *The Poisonous Terrestrial Snakes of Our British Indian Dominions* (4th ed., Bombay: Bombay Natural History Society, 1928)

Wallich, Nathaniel, *Plantae Asiaticae Rariores* (3 vols, London: Teuttel, Würtz & Richter, 1830–32)

Walsh, Cecil, *The Agra Double Murder* (London: Ernest Benn, 1929)

Crime in India (London: Ernest Benn, 1930)

Warden, C. J. H. and L. A. Waddell, *The Non-Baciliar Nature of Abrus-Poison, with Observations on Its Chemical and Physiological Properties* (Calcutta: Bengal Secretariat Press, 1884)

Waring, Edward John, *Pharmacopoeia of India* (London: W. H. Allen, 1868)

Remarks on the Uses of Some of the Bazaar Medicines and Common Medicinal Plants of India (5th ed., London: J. & A. Churchill, 1897)

Watt, George, *A Dictionary of the Economic Products of India* (6 vols, Calcutta: Superintendent, Government Printing, India, 1889–92)

White, A. Denham and Sital Chandra Dutt, 'A Note on the Toxic Symptoms of Organic Arsenic', *IMG* 60 (1925): 464–65

Wilson, H. H., '*Kushta*, or Leprosy, as Known to the Hindus', *TMPSC* 1 (1825): 1–44

'On the Native Practice in Cholera', *TMPSC* 2 (1826): 284–87

Yule, Henry and A. C. Burnell, *Hobson-Jobson: A Glossary of Colloquial Anglo-India Words and Phrases* (2nd ed., London: Routledge & Kegan Paul, 1985)

Secondary sources

Alley, Kelly D., *On the Banks of the Ganga: When Wastewater Meets a Sacred River* (Ann Arbor: University of Michigan Press, 2002)

Anderson, Michael R., 'The Conquest of Smoke: Legislation and Pollution in Colonial Calcutta', in David Arnold and Ramachandra Guha (eds), *Nature, Culture, Imperialism: Essays on the Environmental History of South Asia* (Delhi: Oxford University Press, 1995), 293–335

Anderson, Olive, *Suicide in Victorian and Edwardian England* (Oxford: Clarendon Press, 1987)

Appadurai, Arjun, *The Future as Cultural Fact: Essays on the Global Condition* (London: Verso, 2013)

    (ed.), *The Social Life of Things: Commodities in Cultural Perspective* (Cambridge: Cambridge University Press, 1986)

Arnold, David, 'British India and the Beriberi Problem, 1798–1942', *Medical History* 54 (2010): 295–314

    Colonizing the Body: State Medicine and Epidemic Disease in Nineteenth-Century India (Berkeley: University of California Press, 1993)

    'The "Discovery" of Malnutrition and Diet in Colonial India', *Indian Economic and Social History Review* 31 (1994): 1–26

    'Famine in Peasant Consciousness and Peasant Action: Madras 1876–8', in Ranajit Guha (ed.), *Subaltern Studies III* (Delhi: Oxford University Press, 1984), 62–115

    'Touching the Body: Perspectives on the Indian Plague, 1896–1900', in Ranajit Guha (ed.), *Subaltern Studies V* (Delhi: Oxford University Press, 1987), 55–90

    The Tropics and the Traveling Gaze: India, Landscape, and Science, 1800–1856 (Seattle: Washington University Press, 2006)

Arsenic Policy Support Unit, *Selected Papers on the Social Aspects of Arsenic and Arsenic Mitigation in Bangladesh* (Dhaka: APSU, 2006)

Atkins, P. J., 'White Poison? The Social Consequences of Milk Consumption, 1850–1914', *Social History of Medicine* 5 (1992): 207–27

Baehrel, R., 'La haine de classe en temps d'épidémie', *Annales: Économies, sociétés, civilisations* 7 (1952): 351–60

Ballhatchet, Kenneth, *Race, Sex and Class under the Raj: Imperial Attitudes and Policies and Their Critics, 1793–1905* (London: Weidenfeld & Nicolson, 1980)

Barrett, Ron, *Aghor Medicine: Pollution, Death, and Healing in Northern India* (Berkeley: University of California Press, 2008)

Bartrip, Peter, 'How Green Was My Valence? Environmental Arsenic Poisoning and the Victorian Domestic Ideal', *English Historical Review* 109 (1994): 891–913

    '"A Pennurth of Arsenic for Rat Poison": The Arsenic Act 1851 and the Prevention of Secret Poisoning', *Medical History* 36 (1992): 53–69

Baviskar, Amita, Subir Sinha and Kavita Philip, 'Rethinking Indian Environmentalism: Industrial Pollution in Delhi and Fisheries in Kerala', in Joanne Bauer (ed.), *Forging Environmentalism: Justice, Livelihood, and Contested Environments* (Armonk, NY: M. E. Sharpe, 2006), 189–256

Bayly, C. A., *Empire and Information: Intelligence Gathering and Social Communication in India, 1780–1870* (Cambridge: Cambridge University Press, 1996)

Beals, Alan R., 'Strategies of Resort to Curers in South India', in Charles Leslie (ed.), *Asian Medical Systems: A Comparative Study* (Berkeley: University of California Press, 1976), 184–200

Bertomeu-Sánchez, José Ramón and Agustí Nieto-Galan (eds), *Chemistry, Medicine and Crime: Mateu J. B. Orfila (1787–1853)* (Sagamore Beach: Science History Publications, 2006)

Bhabha, Homi K., *The Location of Culture* (London: Routledge 1994)

Bhargava, K. D. (ed.), *Descriptive List of Mutiny Papers in the National Archives of India, Bhopal* (2 vols, New Delhi: National Archives of India, 1960)

[Bond, Ruskin], *The Best of Ruskin Bond* (New Delhi: Penguin, 1994)

Booth, Martin, *Opium: A History* (New York: Simon & Schuster, 1996)

Bose, Pradip Kumar (ed.), *Health and Society in Bengal: A Selection from Late 19th-Century Bengali Periodicals* (New Delhi: Sage, 2006)

Breeze, Lawrence E., *The British Experience with River Pollution, 1865–1876* (New York: Peter Lang, 1993)

Briggs, David, 'Environmental Pollution and the Global Burden of Disease', in David J. Briggs, Michael Joffe and Paul Elliott (eds), *Impact of Environmental Pollution on Health: Balancing Risk* (Oxford: Oxford University Press, 2003), 1–24

Broughton, Edward, 'The Bhopal Disaster and Its Aftermath: A Review', *Environmental Health* 4 (2005), www.ehjournal.net/content/4/1/6.

Brown, Mark, 'Ethnology and Colonial Administration in Nineteenth-Century British India: The Question of Native Crime and Criminality', *British Journal for the History of Science* 36 (2003): 201–19

Penal Power and Colonial Rule (Abingdon: Routledge, 2014)

Buell, Lawrence, *The Environmental Imagination: Thoreau, Nature Writing, and the Formation of American Culture* (Cambridge: Belknap Press, 1995)

'Toxic Discourse', *Critical Inquiry* 24 (1998): 639–65

Burney, Ian, *Poison, Detection, and the Victorian Imagination* (Manchester: Manchester University Press, 2006)

'A Poisoning of No Substance: The Trials of Medico-Legal Proof in Mid-Victorian England', *Journal of British Studies* 38 (1999): 9–92

Calder, Angus, *Gods, Mongrels and Demons* (London: Bloomsbury, 2003)

Calhoun, Craig (ed.), *Habermas and the Public Sphere* (Cambridge: MIT Press, 1992)

Carson, Rachel, *Silent Spring* (London: Hamish Hamilton, 1963)

Carter, K. Codell, 'The Germ Theory, Beriberi, and the Deficiency Theory of Disease', *Medical History* 21 (1977): 119–36

Centre for Science and Environment, *The State of India's Environment, 1982: A Citizens' Report* (New Delhi: Centre for Science and Environment, 1982)

The State of India's Environment, 1984–85: The Second Citizens' Report (New Delhi: Centre for Science and Environment, 1985)

Chakrabarti, Malabika, *The Famine of 1896–1897 in Bengal: Availability or Entitlement Crisis?* (Hyderabad: Orient Longman, 2004)

Chakrabarti, Pratik, *Bacteriology in British India: Laboratory Medicine and the Tropics* (Rochester: University of Rochester Press, 2012)

Materials and Medicine: Trade, Conquest and Therapeutics in the Eighteenth Century (Manchester: Manchester University Press, 2010)

Western Science in Modern India: Metropolitan Methods, Colonial Practices (Ranikhet: Permanent Black, 2004)

Chakravarty, Papia, *Hindu Response to Nationalist Ferment* (Calcutta: Subarnarekha, 1992)

Chandavarkar, Raj, 'Plague, Panic and Epidemic Politics in India, 1896–1914', in Terence Ranger and Paul Slack (eds), *Epidemics and Ideas: Essays on the*

Historical Perception of Pestilence (Cambridge: Cambridge University Press, 1992), 203–40

Chase, Karen and Michael Levenson, *The Spectacle of Intimacy: A Public Life for the Victorian Family* (Princeton: Princeton University Press, 2000)

Chatterjee, Partha, *The Nation and Its Fragments: Colonial and Postcolonial Histories* (Princeton: Princeton University Press, 1993)

  A Princely Imposter? The Strange and Universal History of the Kumar of Bhawal (Princeton: Princeton University Press, 2002)

  (ed.), *Ranajit Guha: The Small Voice of History: Collected Essays* (Ranikhet: Permanent Black, 2009)

Chaturvedi, G. N., S. K. Tiwari and N. P. Rai, 'Medicinal Use of Opium and Cannabis in Medieval India', *Indian Journal of History of Science* 16 (1981): 31–35

Choudhury, Deep Kanta Lahiri, '"Beyond the Reach of Monkeys and Men"? O'Shaughnessy and the Telegraph in India, c. 1836–56', *Indian Economic and Social History Review* 37 (2000): 331–59

Chowdhury, Indira, 'Delivering the "Murdered Child": Infanticide, Abortion, and Contraception in Colonial India', in Deepak Kumar and Raj Sekhar Basu (eds), *Medical Encounters in British India* (New Delhi: Oxford University Press, 2013), 275–98

Clark, Michael and Catherine Crawford (eds), *Legal Medicine as History* (Cambridge: Cambridge University Press, 1994)

Cohen, Stanley, *Folk Devils and Moral Panics: The Creation of the Mods and Rockers* (3rd ed., London: Routledge, 2002)

Collingham, E. M., *Imperial Bodies: The Physical Experience of the Raj, c. 1800–1947* (Cambridge: Polity Press, 2001)

Copland, I. F. S., 'The Baroda Crisis of 1873–77: A Study in Governmental Rivalry', *Modern Asian Studies* 2 (1968): 97–123

Cronon, William, 'Foreword: The Pain of a Poisoned World', in Brett L. Walker, *Toxic Archipelago*, ix–xii

Dalrymple, William, *The Last Mughal: The Fall of a Dynasty, Delhi, 1857* (London: Bloomsbury, 2006)

Delaporte, François, *Disease and Civilization: The Cholera in Paris, 1832* (Cambridge: MIT Press, 1986)

Derks, Hans, *History of the Opium Problem: The Assault on the East, ca. 1600–1950* (Leiden: Brill, 2012)

Derrida, Jacques, *Dissemination* (London: Athlone Press, 1981)

Desmond, Ray, *The European Discovery of the Indian Flora* (Oxford: Oxford University Press, 1992)

Dirks, Nicholas B., *Castes of Mind: Colonialism and the Making of Modern India* (Princeton: Princeton University Press, 2001)

Doniger, Wendy, *The Hindus: An Alternative History* (New York: Penguin, 2009)

Dossal, Mariam, 'Henry Conybeare and the Politics of Centralised Water Supply in Mid-Nineteenth Century Bombay', *Indian Economic and Social History Review* 25 (1988): 79–96

Douglas, Mary, *Purity and Danger: An Analysis of Concepts of Pollution and Taboo* (London: Routledge, 2002)

Dumont, Louis, *Homo Hierarchicus: The Caste System and Its Implications* (revised ed., Chicago: University of Chicago Press, 1979)

Dutta, Achintya Kumar, 'Medical Research and Control of Disease: Kala-Azar in British India', in Biswamoy Pati and Mark Harrison (eds), *Society, Medicine, and Politics: Colonial India, 1850–1940s* (London: Routledge, 2009), 93–112

Eckerman, Ingrid, *The Bhopal Saga: Causes and Consequences of the World's Largest Industrial Disaster* (Hyderabad: Universities Press, 2005)

Emdad-ul Haq, M., *Drugs in South Asia: From the Opium Trade to the Present Day* (Basingstoke: Palgrave, 2000)

Essig, Mark, 'Poison Murder and Expert Testimony: Doubting the Physician in Late Nineteenth-Century America', *Yale Journal of Law and the Humanities* 14 (2002): 177–210

Everest, Larry, *Behind the Poison Cloud: Union Carbide's Bhopal Massacre* (Chicago: Banner Press, 1986)

Farmer, B. H., *Agricultural Colonization in India since Independence* (London: Oxford University Press, 1974)

Farooq, S. H., D. Chandrasekharam, Z. Berner, S. Norra and D. Stüben, 'Influence of Traditional Agricultural Practices on Mobilization of Arsenic from Sediments to Groundwater in Bengal Delta', *Water Research* 44 (2010): 5575

Fischer-Tiné, Harald and Jana Tschurenev (eds), *A History of Alcohol and Drugs in Modern South Asia: Intoxicating Affairs* (London: Routledge, 2014)

Fisher, Michael H., *Indirect Rule in India: Residents and the Residency System, 1764–1858* (Delhi: Oxford University Press, 1991)

Forbes, Geraldine, 'Managing Midwifery in India', in Dagmar Engels and Shula Marks (eds), *Contesting Colonial Hegemony: State and Society in Africa and India* (London: Academic Press, 1994), 152–72

Foucault, Michel, *The Order of Things: An Archaeology of the Human Sciences* (London: Tavistock Publications, 1970)

Freitag, Sandria B. (ed.), 'Aspects of the Public in Colonial India', special issue, *South Asia* 14 (1991)

Gajalakshmi, Vendham and Richard Peto, 'Suicide Rates in Rural Tamil Nadu, South India', *International Journal of Epidemiology* 36 (2007): 203–07

Ganesan, Uma, 'Medicine and Modernity: The Ayurvedic Revival Movement in India, 1885–1947', *Studies on Asia* 4 (2010): 108–31

Ghosh, Durba, *Sex and the Family in Colonial India: The Making of Empire* (Cambridge: Cambridge University Press, 2006)

Golinski, Jan, *Making Natural Knowledge: Constructivism and the History of Science* (Cambridge: Cambridge University Press, 1998)

Gorman, Mel, 'Sir William Brooke O'Shaughnessy, F.R.S. (1809–1889), Anglo-Indian Forensic Chemist', *Notes and Records of the Royal Society of London* 39 (1984): 51–64

Grey, Daniel J. R., 'Creating the "Problem Hindu": *Sati*, Thuggee and Female Infanticide in India, 1800–860', *Gender and History* 25 (2013): 498–510

Grimshaw, Allen D., 'The Anglo-Indian Community: The Integration of a Marginal Group', *Journal of Asian Studies* 18 (1959): 227–40

Guha, Ramachandra and J. Martinez-Alier, *Varieties of Environmentalism: Essays North and South* (London: Verso, 1997)

Guha, Ranajit, 'Chandra's Death', in Partha Chatterjee (ed.), *Ranajit Guha: The Small Voice of History: Collected Essays* (Ranikhet: Permanent Black, 2009), 271–303

'Not at Home in Empire', in Partha Chatterjee (ed.), *Ranajit Guha: The Small Voice of History: Collected Essays* (Ranikhet: Permanent Black, 2009), 441–54

Guha, Supriya, 'The Unwanted Pregnancy in Colonial Bengal', *Indian Economic and Social History Review* 33 (1996): 403–36

Gunnell, David and Michael Eddleston, 'Suicide by Intentional Ingestion of Pesticides', *International Journal of Epidemiology* 32 (2003): 902–09

Gupta, Charu, *Sexuality, Obscenity, Community: Women, Muslims, and the Hindu Public in Colonial India* (Delhi: Permanent Black, 2001)

Haberman, David L., *River of Love in an Age of Pollution: The Yamuna River of Northern India* (Berkeley: University of California Press, 2006)

Harewood, Earl of, (ed.), *Kobbé's Complete Opera Book* (London: Putnam, 1981)

Harper, Edward B., 'Ritual Pollution as an Integrator of Caste and Religion', *Journal of Asian Studies* 23 (1964): 151–97

Harrison, Mark, *Climates and Constitutions: Health, Race, Environment and British Imperialism in India, 1600–1850* (New Delhi: Oxford University Press, 1999)

Medicine in an Age of Commerce and Empire: Britain and Its Tropical Colonies, 1660–1830 (Oxford: Oxford University Press, 2010)

Public Health in British India: Anglo-Indian Preventive Medicine, 1859–1914 (Cambridge: Cambridge University Press, 1994)

'Racial Pathologies: Morbid Anatomy in British India, 1770–1850', in Biswamoy Pati and Mark Harrison (eds), *The Social History of Health and Medicine in Colonial India* (London: Routledge, 2009), 173–94

Hawgood, Barbara J., 'The Life and Viper of Dr Patrick Russell: Physician and Naturalist', *Toxicon* 32 (1994): 1295–304

'Sir Joseph Fayrer: Snakebite and Mortality in British India', *Toxicon* 34 (1996): 171–82

Heath, Deana, *Purifying Empire: Obscenity and the Politics of Moral Regulation in Britain, India and Australia* (Cambridge: Cambridge University Press, 2010)

Hill, Marquita K., *Understanding Environmental Pollution* (Cambridge: Cambridge University Press, 1997)

Hume, John C., 'Rival Traditions: Western Medicine and Yunan-i Tibb in the Punjab, 1849–1889', *Bulletin of the History of Medicine* 51 (1977): 214–31

Hussain, Naila, *Poisoned Lives: The Effects of Cotton Pesticides* (Lahore: Shirkat Gah, 1999)

Ions, Veronica, *Indian Mythology* (London: Newnes Books, 1983)

Jackson, Mark, '"Divine Stramonium": The Rise and Fall of Smoking for Asthma', *Medical History* 54 (2010): 171–94

Jackson, Peter, *The Delhi Sultanate: A Political and Military History* (Cambridge: Cambridge University Press, 1999)

Jain, S. K. and S. K. Borthakur, 'Solanaceae in Indian Tradition, Folklore, and Medicine', in William G. D'Arcy (ed.), *Solanaceae: Biology and Systematics* (New York: Columbia University Press, 1986), 577–83

Jeffrey, Robin, *The Decline of Nayar Dominance: Society and Politics in Travancore, 1847–1908* (New York: Holmes & Meier, 1976)

Johnson, Ryan and Amna Khalid (eds), *Public Health in the British Empire: Intermediaries, Subordinates, and the Practice of Public Health, 1850–1960* (New York: Routledge, 2013)

Jolly, Julius, *Indian Medicine* (New Delhi: Munshiram Manoharlal, 1977)

Joshi, Sanjay, *Fractured Modernity: Making of a Middle-Class in Colonial North India* (Delhi: Oxford University Press, 2001)

Karan, P. P. and W. A. Bladen, 'Geographical Aspects of Environmental Pollution in India', *Geoforum* 7 (1976): 51–57

Kasturi, Malavika, 'Law and Crime in India: British Policy and the Female Infanticide Act of 1870', *Indian Journal of Gender Studies* 1 (1994): 169–93

'Taming the "Dangerous" Rajput: Family, Marriage and Female Infanticide in Nineteenth-Century Colonial North India', in Harald Fischer-Tiné and Michael Mann (eds), *Colonialism as Civilizing Mission: Cultural Ideology in British India* (London: Anthem Press, 2004), 117–40

Keen, Caroline, *Princely India and the British: Political Development and the Operation of Empire* (London: I. B. Tauris, 2012)

Kinkela, David, *DDT and the American Century: Global Health, Environmental Politics, and the Pesticide That Changed the World* (Chapel Hill: University of North Carolina Press, 2011)

Klein, Ira, 'Death in India, 1871–1921', *Journal of Asian Studies* 22 (1973): 639–59

Kolsky, Elizabeth, *Colonial Justice in British India: White Violence and the Rule of Law* (Cambridge: Cambridge University Press, 2010)

Kopytoff, Igor, 'The Cultural Biography of Things: Commoditization as Process', in Arjun Appadurai (ed.), *Social Life of Things*, 64–91

Koselleck, Reinhart, *Futures Past: On the Semantics of Historical Time* (Cambridge: MIT Press, 1985)

The Practice of Conceptual History: Timing History, Spacing Concepts (Stanford: Stanford University Press, 2002)

Kumar, Anil, *Medicine and the Raj: British Medical Policy in India, 1835–1911* (New Delhi: Sage, 1998)

Lang, Sean, '"Drop the Demon *Dai*", Maternal Mortality and the State in Colonial Madras, 1840–1875', *Social History of Medicine* 18 (2005): 357–78

Lefebvre, George, *The Great Fear of 1789: Rural Panic in Revolutionary France* (New York: Pantheon Books, 1973)

Leslie, Charles, 'The Ambiguities of Medical Revivalism in Modern India', in Leslie (ed.), *Asian Medical Systems*, 356–67

(ed.), *Asian Medical Systems: A Comparative Study* (Berkeley: University of California Press, 1976)

Levey, Martin, 'Medieval Arabic Toxicology: The *Book of Poisons* of Ibn Wahshiya and Its Relation to Early Indian and Greek Texts', *Transactions of the American Philosophical Society* 56 (1966): 1–130

Ludden, David, 'India's Development Regime', in Nicholas B. Dirks (ed.), *Colonialism and Culture* (Ann Arbor: University of Michigan Press, 1992), 247–87

Mann, Harold H., 'The Untouchables of an Indian City', in Daniel Thorner (ed.), *Harold H. Mann: The Social Framework of Agriculture: India, Middle East, England* (Bombay: Vora, 1967), 175–91

Mann, Michael, 'Delhi's Belly: On the Management of Water, Sewage and Excreta in a Changing Urban Environment during the Nineteenth Century', *Studies in History* 23 (2007): 1–31

Manor, James, *Power, Poverty and Poison: Disaster and Response in an Indian City* (New Delhi: Sage, 1993)

Marco, Gino J., Robert M. Hollingworth and William Durham (eds), *Silent Spring Revisited* (Washington: American Chemical Society, 1987)

Maskiell, Michelle and Adrienne Mayor, 'Killer Khilats, Part 1: Legends of Poisoned "Robes of Honour" in India', *Folklore* 112 (2001): 23–45

and Adrienne Mayor, 'Killer Khilats, Part 2: Imperial Collecting of Poison Dress Legends in India', *Folklore* 112 (2001): 163–82

Mawdsley, Emma, 'India's Middle Classes and the Environment', *Development and Change* 35 (2004): 79–103

McFarlane, Colin, 'Governing the Contaminated City: Infrastructure and Sanitation in Colonial and Post-Colonial Bombay', *International Journal of Urban and Regional Research* 32 (2008): 415–35

Meharg, Andrew A., *Venomous Earth: How Arsenic Caused the World's Worst Mass Poisoning* (Basingstoke: Macmillan, 2005)

Mills, James H, *Cannabis Britannica: Empire, Trade, and Prohibition, 1800–1928* (Oxford: Oxford University Press, 2003)

'Drugs, Consumption, and Supply in Asia: The Case of Cocaine in Colonial India, c. 1900–c.1930', *Journal of Asian Studies* 66 (2007): 345–62

Mishra, Saurabh, *Beastly Encounters of the Raj: Livelihoods, Livestock and Veterinary Health in North India, 1790–1920* (Manchester: Manchester University Press, 2015)

'Of Poisoners, Tanners and the British Raj: Redefining Chamar Identity in Colonial North India, 1850–90', *Indian Economic and Social History Review* 48 (2011): 317–38

Mukherjee, Suroopa, *Surviving Bhopal: Dancing Bodies, Written Texts, and Oral Testimonials of Women in the Wake of an Industrial Disaster* (New York: Palgrave Macmillan, 2010)

Mukherjee, Upamanyu Pablo, *Postcolonial Environments: Nature, Culture and the Contemporary Indian Novel in English* (Basingstoke: Palgrave Macmillan, 2010)

Naraindas, Harish, 'Poisons, Putrescence and the Weather: A Genealogy of the Advent of Tropical Medicine', *Contributions to Indian Sociology* 30 (1996): 1–35

Newman, David L., (ed.), *The Sultan's Sex Potions: Arab Aphrodisiacs in the Middle Ages* (London: Saqi Books, 2014)

Newman, Richard, 'Early British Encounters with the Indian Opium Eater', in James H. Mills and Patricia Barton (eds), *Drugs and Empires: Essays in*

Modern Imperialism and Intoxication, c.1500–c.1930 (Basingstoke: Palgrave Macmillan, 2007), 57–72

Nigam, Sanjay, 'Disciplining and Policing the "Criminals by Birth", Part 1: The Making of a Colonial Stereotype – The Criminal Tribes and Castes of North India', *Indian Economic and Social History Review* 27 (1990): 131–64

Nixon, Rob, *Slow Violence and the Environmentalism of the Poor* (Cambridge: Harvard University Press, 2011)

Noorani, A. O., 'The Citizens' Environment', *Economic and Political Weekly* 19, nos 51–52 (22–29 December 1984): 2150.

Obringer, Frédéric, *L'Aconit et L'Orpiment: Drogues et Poisons en Chine Ancienne et Médiévale* (Paris: Librairie Arthème Fayard, 1997)

Owen, David Edward, *British Opium Policy in China and India* (New Haven: Yale University Press, 1934)

Palsetia, Jesse S., 'Mad Dogs and Parsis: The Bombay Dog Riots of 1832', *Journal of the Royal Asiatic Society* 11 (2001): 13–30

Panda, Ashok Kumar and Saroj Kumar Debnath, 'Overdose Effect of Aconite-Containing Ayurvedic Medicine', *International Journal of Ayurveda Research* 1 (2010): 183–86

Pande, Ishita, *Medicine, Race and Liberalism in British Bengal: Symptoms of Empire* (Abingdon: Routledge, 2009)

Pandey, Gyanendra, *The Construction of Communalism in Colonial North India* (Delhi: Oxford University Press, 1990)

Panikkar, K. N., 'Indigenous Medicine and Cultural Hegemony: A Study of the Revitalization Movement in Keralam', *Studies in History* 8 (1992): 283–308

Parel, Anthony J. (ed.), *M. K. Gandhi: Hind Swaraj and Other Writings* (Cambridge: Cambridge University Press, 1997)

Pelling, Margaret, *Cholera, Fever and English Medicine, 1825–1865* (Oxford: Oxford University Press, 1978)

Pillai, S. Chidambara Thanu, *Siddha System of Toxicology* (Madras: Siddha Medical Literature Research Centre, 1993)

Procida, Mary, 'Feeding the Imperial Appetite: Imperial Knowledge and Anglo-Indian Discourse', *Journal of Women's History* 15 (2003): 123–49

Ramasubban, Radhika, 'Profit against Safety', *Economic and Political Weekly* 19, nos 51–52 (22–29 December 1984): 2147–50

Ramusack, Barbara N., 'Incident at Nabha: Interaction between Indian States and British Indian Politics', *Journal of Asian Studies* 28 (1969): 563–77

Robb, George, 'Circe in Crinoline: Domestic Poisonings in Victorian England', *Journal of Family History* 22 (1997): 176–90

Robb, Peter, *Sentiment and Self: Richard Blechynden's Calcutta Diaries, 1791–1822* (New Delhi: Oxford University Press, 2011)

Sex and Sensibility: Richard Blechynden's Calcutta Diaries, 1791–1822 (New Delhi: Oxford University Press, 2011)

Roberts, Donald R. and Richard Tren, 'Did Rachel Carson Understand the Importance of DDT in Global Public Health Programs?' in Roger Meiners, Pierre Desrochers and Andrew Morriss (eds), *Silent Spring at 50: The False Crises of Rachel Carson* (Washington: Cato Institute, 2012), 167–99

Roche, Daniel, *A History of Everyday Things: The Birth of Consumption in France,*
1600–1800 (Cambridge: Cambridge University Press, 2000)

Said, Edward W., *Orientalism* (London: Routledge & Kegan Paul, 1978)

Sangar, S. P., 'Intoxicants in Ancient India', *Indian Journal of History of Science* 16
(1981): 204–14

Sarkar, S. C., *Notable Indian Trials* (3rd ed., Calcutta: M. C. Sarkar & Sons, 1962)

Sarkar, Sumit, *Modern India, 1885–1947* (2nd ed., Basingstoke: Macmillan,
1989)

Sarkar, Tanika, *Hindu Wife, Hindu Nation: Community, Religion and Cultural*
Nationalism (London: Hurst, 2001)

Schiebinger, Londa, *Plants and Empire: Colonial Bioprospecting in the Atlantic World*
(Cambridge: Harvard University Press, 2004)

Scott, James C., *Seeing Like a State: How Certain Schemes to Improve the Human*
Condition Have Failed (New Haven: Yale University Press, 1998)

Weapons of the Weak: Everyday Forms of Peasant Resistance (New Haven: Yale
University Press, 1985)

Searle-Chatterjee, Mary, 'The Polluted Identity of Work: A Study of Benares
Sweepers', in Sandra Wallmann (ed.), *Social Anthropology of Work* (London:
Academic Press, 1979), 269–86

Sen, Indrani, '"Cruel, Oriental Despots": Representations in Nineteenth-
Century British Colonial Fiction, 1858–1900', in Waltraud Ernest and
Biswamoy Pati (eds), *India's Princely States: People, Princes and Colonialism*
(London: Routledge, 2007), 30–48

Sen, Satadru, 'The Savage Family: Colonialism and Female Infanticide in
Nineteenth-Century India', *Journal of Women's History* 14 (2002): 53–79

Sen, Sudipta, 'Colonial Aversions and Domestic Desires: Blood, Race, Sex, and
the Decline of Intimacy in Early British India', in Sanjay Srivastava (ed.),
Sexual Sites, Seminal Attitudes: Sexualities, Masculinities and Culture in South
Asia (New Delhi: Sage, 2004), 49–82

Sengoopta, Chandak, *Imprint of the Raj: The Colonial Origins of Fingerprinting and*
Its Voyage to Britain (London: Macmillan, 2003)

Sharan, Awadhendra, *In the City, Out of Place: Nuisance, Pollution, and Dwelling in*
Delhi, c. 1850–2000 (New Delhi: Oxford University Press, 2014)

Sharma, Meenakshi, 'Polluted River or Goddess and Saviour? The Ganga in the
Discourses of Modernity and Hinduism', in Helen Tiffin (ed.), *Five Emus to*
the King of Siam: Environment and Empire (Amsterdam: Rodopi, 2007),
31–50

Shiva, Vandana, 'Reductionist Science as Epistemological Violence', in Ashis
Nandy (ed.), *Science, Hegemony and Violence: A Requiem for Modernity* (New
Delhi: Oxford University Press, 1988)

and Kunwar Jalees, *Farmers Suicides in India* (New Delhi: Research Foundation
for Science, Technology and Ecology, n.d. [c. 2004])

Siegal, Lee, *Sacred and Profane Dimensions of Love in Indian Traditions as*
Exemplified in the Gitagovinda *of Jayadeva* (Delhi: Oxford University Press,
1978)

Singha, Radhika, *A Despotism of Law: Crime and Justice in Early Colonial India*
(Delhi: Oxford University Press, 1998)

Sinha, Mrinalini, *Colonial Masculinity: The 'Manly Englishman' and the 'Effeminate Bengali' in the Late Nineteenth Century* (Manchester: Manchester University Press, 1995)

Smith, Allan H., Elena O. Lingas and Mahfuzar Rahman, 'Contamination of Drinking-Water by Arsenic in Bangladesh: A Public Health Emergency', *Bulletin of the World Health Organization* 78 (2000): 1093–103

Srivastava, R. S., 'Malaria Control Measures in the Tarai Area under the Tarai Colonization Scheme', *IJM* 4 (1950): 151–65

Stoler, Ann Laura, *Carnal Knowledge and Imperial Power: Race and the Intimate in Colonial Rule* (Berkeley: University of California Press, 2002)

'Tense and Tender Ties: The Politics of Comparison in North American History and (Post) Colonial Studies', *Journal of American History* 88 (2001): 829–65

Suleri, Sara, *The Rhetoric of English India* (Chicago: University of Chicago Press, 1992)

Sundar, K. M. Shyam, *Treatment for Poisons in Traditional Medicine* (Madras: Centre for Indian Knowledge Systems, 1996)

Teitelbaum, Emmanuel, 'Was the Indian Labor Movement Ever Co-Opted?', *Critical Asian Studies* 38 (2006): 389–417

Thompson, Kenneth, *Moral Panics* (London: Routledge, 1998)

Tomic, Sacha, 'Alkaloids and Crime in Early Nineteenth-Century France', in José Ramón Bertomeu-Sánchez and Agustí Nieto-Galan (eds), *Chemistry, Medicine, and Crime: Mateu J. B. Orfila (1787–1853) and His Times* (Sagamore Beach: Science History Publications, 2006), 261–92

Trocki, Carl A., *Opium, Empire and the Global Political Economy: A Study of the Asian Opium Trade, 1750–1950* (London: Routledge, 1999)

Vaughan, Megan, *Creating the Creole Island: Slavery in Eighteenth-Century Mauritius* (Durham: Duke University Press, 2005)

Viswanathan, D. K., *The Conquest of Malaria in India: An Indo-American Co-Operative Effort* (Madras: Company Law Institute Press, 1958)

Wagner, Kim A., 'The Deconstructed Stranglers: A Reassessment of Thuggee', *Modern Asian Studies* 38 (2004): 931–63

The Great Fear of 1857: Rumours, Conspiracies and the Making of the Indian Uprising (Oxford: Peter Lang, 2010) 31–63

Stranglers and Bandits: A Historical Anthology of Thuggee (New Delhi: Oxford University Press, 2009)

Thuggee: Banditry and the British in Early Nineteenth-Century India (Basingstoke: Palgrave Macmillan, 2007)

'"Treading upon Fires": The "Mutiny"-Motif and Colonial Anxieties in British India', *Past and Present*, no. 218 (2013): 159–97

Walker, Brett L., *Toxic Archipelago: A History of Industrial Disease in Japan* (Seattle: University of Washington Press, 2010)

Watson, Katherine D., *Forensic Medicine in Western Society: A History* (Oxford: Routledge, 2011)

'Medical and Chemical Expertise in English Trials for Criminal Poisoning, 1750–1914', *Medical History* 50 (2006): 373–90

Poisoned Lives: English Poisoners and Their Victims (London: Hambledon Continuum, 2004)

Whittington-Egan, Molly, *Khaki Mischief: The Agra Murder Case* (London: Souvenir Press, 1990)

Whyte, Susan Reynolds, Sjaak van der Geest and Anita Harden (eds), *Social Lives of Medicine* (Cambridge: Cambridge University Press, 2002)

Wiener, Martin, 'Alice Arden to Bill Sykes: Changing Nightmares of Intimate Violence in England, 1558–1869', *Journal of British Studies* 40 (2001), 184–212

Wilkinson, C. F., 'The Science and Politics of Pesticides', in Gino J. Marco, Robert M. Hollingworth and William Durham (eds), *Silent Spring Revisited* (Washington: American Chemical Society, 1987), 25–46

Wingate, Peter, with Richard Wingate, *The Penguin Medical Encyclopedia* (3rd ed., London: Penguin, 1988)

Winther, Paul C., *Anglo-European Science and the Rhetoric of Empire: Malaria, Opium, and British Rule in India, 1756–1895* (Lanham: Lexington Books, 2003)

Whorton, James, C., *The Arsenic Century: How Victorian Britain Was Poisoned at Home, Work, and Play* (Oxford: Oxford University Press, 2010)

Wootton, David, *Bad Medicine: Doctors Doing Harm since Hippocrates* (Oxford: Oxford University Press, 2006)

Worboys, Michael, *Spreading Germs: Disease Theories and Medical Practice in Britain, 1865–1900* (Cambridge: Cambridge University Press, 2000)

Wujastyk, Dominik, *The Roots of Ayurveda: Selections from Sanskrit Medical Writings* (New Delhi: Penguin, 2001)

Yang, Anand. A., 'A Conversation of Rumors: The Language of Popular *Mentalités* in Late Nineteenth-Century Colonial India', *Journal of Social History* 20 (1987): 485–505

Yangwen, Zheng, *The Social Life of Opium in China* (Cambridge: Cambridge University Press, 2005)

Zachariah, Benjamin, *Developing India: An Intellectual and Social History, c. 1930–1950* (New Delhi: Oxford University Press, 2005)

Ziegler, Philip, *The Black Death* (Harmondsworth: Penguin, 1982)

Zimmer, Henry R., *Hindu Medicine* (Baltimore: Johns Hopkins University Press, 1948)

Zimmermann, Francis, *The Jungle and the Aroma of Meats: An Ecological Theme in Hindu Medicine* (Berkeley: University of California Press, 1987)

Index